T0271928

POLYMER NETWORKS '91

Polymer Networks '91

Proceedings of the International Conference, 21-26 April 1991,
Moscow, Russia

Editors:
K. Dušek
Czechoslovak Academy of Sciences, Prague,
Czechoslovakia

and

S.I. Kuchanov
Moscow State University, Moscow, Russia

///VSP///

1992
Utrecht, The Netherlands
Tokyo, Japan

VSP BV
P.O.Box 346
3700 AH Zeist
The Netherlands

©VSP BV 1992

First published in 1992

ISBN 90-6764-145-6

CIP-DATA KONINKLIJKE BIBLIOTHEEK, DEN HAAG

Polymer networks '91: proceedings of the international conference,
21-26 April 1991, Moscow, USSR / eds.: K. Dusek and S.I. Kuchanov
- Utrecht: VSP.
ISBN 90-6764-145-6 bound
NUGI 813
Subject heading: polymer networks

Printed in Great Britain by Bookcraft (Bath) Ltd., Midsomer Norton.

CONTENTS

Preface

The present collection comprises chapters which are an extended version of plenary lectures delivered by well-known scientists during the International Conference 'Polymer Networks - '91' held in Moscow (21-26 April, 1991).

When investigating Polymer Networks an explorer usually faces quite different problems connected with the peculiarities of their synthesis, structure and properties. The diversity of these problems stipulates the involvement for their successful solution of the scientists specializing in Polymer Organic Chemistry, Physical Chemistry, Theoretical as well as Experimental Physics, Materials Science and so on. The main target of the meeting was to establish contacts and to stimulate the exchange of ideas and experience among participants engaged in the investigation of various features of Polymer Networks. The vast majority of the communications presented were of indisputable high scientific level and numerous valuable results have been reported at the Conference.

The present collection aims to bring these useful results to the attention of the wide range of scientists dealing with polymer networks. In this book the papers on Covalent Networks written by authoritative experts from the UK, Russia, Germany, France, Denmark, Czechoslovakia, Poland and Hungary are presented. Another part of the Proceedings of the Conference 'Polymer Networks - '91' treating the Thermoreversible, Interpenetrating and Charged Polymer Networks is to be published elsewhere.

I am highly indebted to VSP International Science Publishers, for their cooperation in preparing this collection for publication.

On behalf of the Editors
S.I. Kuchanov
Moscow State University
Moscow
Russia

Polymer Networks '91 pp. 1-6
Dosek and Kuchanov (Eds)
© VSP 1992

Formation processes, structure and properties of polymer networks

K. Dušek

Institute of Macromolecular Chemistry, Czechoslovak Academy of Sciences, 162 06 Prague, Czech and Slovak Federative Republic

ABSTRACT

Current network formation theories and their applicability to real chemical systems are briefly reviewed. Attention is paid to the effect of reactivities on network formation and to long-range correlations arising from the time sequences of bonds and interaction in space, particularly to their simulation. The applications of the network formation theories to various systems of technological importance are also discussed.

INTRODUCTION

The structure of polymer networks may be very different. Some of them are homogeneous and the crosslinks are distributed in space more or less at random. Some of the networks are less homogeneous and one can find in them regions of higher and lower crosslinking density. Sometimes, micronetworks (microgels) are formed first and then combined into macronetworks.

Loose networks of vulcanized rubber and very dense silica networks can serve as examples of homogeneous networks. Network formation by free-radical copolymerizations of divinyl monomers proceeds via formation of microgel-like micronetworks [1]. The macronetworks of silica can be composed either of star-shaped or compact spherical micronetworks depending on the conditions of preparation [2].

Therefore, the structures of networks, even of comparable crosslinking density, can be very manifold and in order to understand the networks structure one has to understand the network formation process. The bridge relating structure to network formation is a network formation theory. An overview of the recent development in network formation theories and applications can be found elsewhere [3].

NETWORK FORMATION THEORIES

The growth of chemical structures in a network formation process is determined primarily by the chemical rules of making bonds between the starting components. These rules are given by the chemical mechanisms and kinetics (reactivities) of groups. The reactivities (rate constants) are usually dependent only on their neighborhood, i.e. on the state of the building unit of which they are a part. The state of a building unit is defined by the types and numbers of bonds the unit is bound to neigboring units. Thus, the mechanism and reactivity factors are usually short range.

However, these short-range *reactivity and reaction mechanism correlations* can induce long-range correlations in the structure which stem from the fact that the system can remember its history. These are called *time correlations*. An initiated polyaddition can serve as an example: the distribution of chain lengths is dependent on the relative rates of initiation and monomer addition. The distribution is different from that obtained by random combination of constituent units. The time correlations are independent of the dimensionality of space.

There also exist *physical interactions* depending on *dimensionality of space*. Among them, *cyclization* is determined by the probability that two groups already connected by one or more sequences of bonds meet in space and form a bond.

The other spatial factors are *excluded volume effects* and *diffusion control*. The excluded volume means that the apparent reactivity of a pair of groups is determined by the probability that they meet in space. This probability may be limited by thermodynamic or steric effects and depends on the size and geometry of the reacting molecules and position of the group. Similar effect has diffusion control of meeting of the two reacting groups. Diffusion limited processes may give rise to spatial fluctuations in the density of groups.

All these correlations affect not only the structure growth (kinetics) but also the structure itself. The existing network formation theories do or do not take these correlations into account or they simulate them using various approximations. The theories can be grouped in two major categories:

1. Models not directly associated with dimensionality of space,

2. Computer simulations of structure growth in n-dimensional space.

These theories have been discussed in several reviews [4] - [8]. Within group 1, structures can be generated either by (a) statistical methods from building units or (b) kinetic differential equations (coagulation equation).

Statistical Methods

Statistical methods (Flory-Stockmayer theory, theory of branching processes (cascade theory), Miller-Macosko recursive method, etc.) work with building units in different reaction states and structures are generated by random combination of corresponding reacted functional groups. The method is rigorous for equilibrium controlled reactions. Often but not always it is a good approximation for kinetically controlled reactions.

Kinetic Methods

In contrast to statistical theories, the kinetic or coagulation theories preserve the integrity of structures developed during network formation intact. The development of all species is described by an (infinite) set of kinetic differential equations [9]. The resulting distribution can be obtained analytically only in the simplest cases (random reactions), in the other cases moments of distributions can be obtained numerically. The set of differential equations can be also solved by Monte-Carlo methods. The application of the kinetic theory has the disadvantage that the theory considers the gel only as one (giant) molecule and cannot generate structural parts characteristic for the gel (elastically active chains, dangling chains, etc.).

Combinations

This disadvantage can be removed for some kinetically controlled reactions by combination of the statistical and kinetic methods [10], [11]. The strategy employs the following fact: connections between groups of independent reactivity do not transfer information and can be split and again reformed at random. This is true for many starting components. By splitting these connections and labelling the points of cut, the functionality of the units is lowered. Then kinetic method is applied to this new system of lower functionality and a new distribution of oligomers is obtained. Usually, this new distribution is still finite. The generated oligomers still carry the labels. In the last step, network is formed by random combination of the labelled points of cut.

Simulations in Space

The most widely used technique of structure growth simulation is percolation. Percolation is usually understood to be carried out on lattices [4], however off-lattice simulations are also used. Lattice percolations uses some implicit assumptions about the behavior of the system: (a) dependence of structural development on the lattice type and (b) complete rigidity of the system without conformational rearrangements and diffusion. Percolation techniques are at the moment not very suitable general methods for correlations between structure and structure growth parameters but they seem to be useful for examining the structure development near the gel point [2]. For some special systems, where the structure growth is much faster than diffusion, they seem to offer reasonable predictions (e.g. kinetic (initiated) percolation for free-radical polymerizations) [13].

APPLICATIONS OF BRANCHING THEORIES

The branching theories have been applied to a number of model systems as well as to systems of technological importance. The purpose of studies of model systems was to test the branching theories as well as the molecular theories of some structure-sensitive properties of systems undergoing crosslinking and fully crosslinked systems. Particularly, the equilibrium rubber elasticity was addressed.

In general, the applicability of the relevant branching theories was confirmed. The parameters studied were the molar mass distributions or averages of the branched polymers before the gel point and their scattering behavior, critical conversion at the gel point, development of the gel fraction and increase of the equilibrium elastic modulus proportional to the concentration of elastically active network chains. The agreement was generally good if a theory consistent with the chemical mechanism and kinetics was chosen and the effect of possible physical interactions taken into account. Problems still exist with the interpretation of rubber elastic behavior due to a number of competing theories and a certain ambiguity of experimental data. However, the conclusions should not be too pesimistic because predictions by the alternative theories may differ by several tens of percents, whereas the correlation exists over several orders of magnitude in crosslinking density.

There exist important properties of crosslinked polymers where the relation to the structure is not so straightforward and molecular theories are still to be developed. These are for example rheological and viscoclastic properties during crosslinking, ultimate properties, thermal properties, etc. However, the information supplied by network formation theories will represent a necessary input data for structure-property relations.

The major success of the network formation theories can be seen in their applications to complicated systems of technological importance. Below, only some of the applications are listed with some general references:

Curing of epoxy resins with various curing agents [5],

Formation of polyurethane networks [6],[14],

Crosslinking and degradation [15],

Crosslinking of various degree-of-polymerization and functionality distributions [16],

Multistage processes in which the final network is formed in several stages [17],

Polyvinyl monomers copolymerization with strong cyclization [13].

CONCLUSIONS

The existing network formation theories can take into account various features of the chemical mechanism and kinetics and can approximate more or less successfully the long-range spatial correlations resulting in cyclization, excluded volume and diffusion controls of the structure growth. Many complex network formation processes important for technologies and applications can be dealt with by the existing theories. Their application helps the chemist and technologist to understand each other.

ACKNOWLEDGEMENT. A partial support by the Grant Agency of Czechoslovak Academy of Sciences is appreciated.

REFERENCES

[1] K. Dušek, in: Developments in Polymerisation. 3., R.N. Haward (Ed.), p.143, Applied Science Publ., Barking (1982).

[2] J.K. Klems and D. Posselt, in: Random Fluctuations and Pattern Growth, H.E. Stanley and N. Ostrowski (Eds.), p. 7, Kleuwe Acad. Press, Dordrecht (1988).

[3] K. Dušek, Rec. Trav. Chim. Pays-Bas, in press.

[4] D. Stauffer, A. Coniglio and M. Adam, Adv. Polym. Sci. 44, 103 (1981).

[5] K. Dušek, Adv. Polym. Sci. 78, 1 (1986).

[6] K. Dušek, in: Telechelic Polymers, J. Goethals (Ed.), p. 289, CRC Press, Boca Raton (1988).

[7] W. Burchard, Adv. Polym. Sci. 48, 1 (1982)

[8] S.I. Kuchanov, S.V. Korolev and S.V. Panyukov, Adv. Chem. Phys. 72, 115 (1988).

[9] Kinetics of Aggregation and Gelation, F. Family and D.P. Landau (Eds.), Elsevier (1984).

[10] K. Dušek, Brit. Polym. J. 17, 185 (1985).

[11] K. Dušek and J. Šomvársky, Polym. Bull. 13 , 313 (1985).

[12] M. Adam, Makromol. Chem., Macromol. Symp., 45, 1 (1991)

[13] H. M. J. Boots, in: Integration of Polymer Science and Technology, L.A. Kleintjens and P.J. Lemstra (Eds.), p.204. Elsevier(1986)

[14] K. Dušek, M. Špírková and I. Havlíček, Macromolecules 23. 1774 (1990).

[15] D.R. Miller and C.W. Macosko, J. Polym. Sci., Polym. Phys. Ed., 26, 1 (1988).

[16] K. Dušek and M. Demjanenko, Rad. Phys. Chem. 28, 479 (1986).

[17] B.J.R. Scholtens, G.P.J.M. Tiemersma-Thoone, K. Dušek and M. Gordon, J. Polym. Sci., Polym. Phys. Ed., 29, 463 (1991).

Polymer Networks '91 pp. 7-24
Dosek and Kuchanov (Eds)
© VSP 1992

Comparative analysis of the processes of polymer networks formation via polycondensation and polymerization

B.A. Rozenberg and V.I. Irzhak

Institute of Chemical Physics, Russian Academy of Sciences, 142432 Chernogolovka Moscow region, Russia

ABSTRACT

Some comparative characteristics of the polycondensation and polymerization kinetic features are discussed. It was shown that just kinetic features of the polymer formation method determine not only the characteristics of the molecular structure but topological and supermolecular ones of forming polymer as well.

INTRODUCTION

The processes of the synthesis of the network as well as linear polymers can be divided into two types: polycondensation and polymerization by means of any known mechanisms. Kinetic regularities of these processes of polymer formation have been thoroughly studied and this knowledge is the basis of a polymer synthesis. Nevertheless, there are difficulties to giving a proper answer on the question, whether two polymer samples with the same molecular structure prepared by polycondensation and polymerization methods posses the same properties or not. Until now there is no solution to the problem considering the connection of polymer structure characteristics (topological and supermolecular organization), finally determining it properties, and kinetic features of its formation. The given paper is devoted to discussion of some aspects of this many-sided problem.

SOME DEFINITIONS

The main difference between polycondensation and polymerization processes from the kinetic point of view consists in the way of polymer chain formation [1].

The polycondensation process proceeds according to general kinetic equation:

$$R_{i,j} + R_{m,n} \longrightarrow R_{(i+m),(j+n-2)} \tag{1}$$

In this case the functional groups of chains can react with each other. Therefore, a chain is formed by assembly of separate fragments.

The polymerization process can be described in the following way:

$$R^*_{i,j,k} + M_{1,n} \longrightarrow R^*_{(i+1),(j+n-1),k} \tag{2}$$

i.e. the process of chain assembly is the result of a successive addition of single units to the active propagating chain $R^*_{i,j,k}$. Here R are polymer chains, M is monomer. First index characterizes the number of monomer units, the second one - a number of functional groups and the third one - a number of active centers in a chain. Reactions (1) and (2) can be irreversible or reversible. For simplification we will consider below only irreversible reactions.

The definition of these processes to our mind completely exhausts the kinetic aspect of the problem. However, there are some different definitions [2,3]. It is worth making two remarks. The first one: in contrast to the widely spread definition of polycondensation as the process of polymer formation that was accompanied by elimination of low molecular compounds, this feature of polycondensation can be ignored in the kinetic definition used here. The second one: the polyaddition reactions without any elimination of low molecular substances (like formation of epoxy-amine or polyurethane polymers and so on) in used classification are also considered as polycondensation reactions. We also must note that the formation of network polymers by crosslinking of already prepared polymers can be formally described as polycondensation or polymerization processes [1].

Some molecular and topological characteristics (distribution of the comonomer or stereoisomer units in the chain, molecular weight distribution, sol-fraction, etc.) of polymers formed via polycondensation and polymerization were the subject of intensive investigation [2-6]. Up to now these aspects have been studied sufficiently well. In the same time influence of the kinetic peculiarities of a polymer formation process on the morphology of polymer formed is not solved yet, in spite of intensive discussion in the literature [1,7-10]. It was shown that the network polymers in some cases are formed via the stage of microgel formation [7] resulting in the inhomogeneous distribution of polymer in space. This mechanism has become very popular and widely used for explanation of a network polymers properties irrespective of their origin. Below we will show that such an approach is wrong and the morphology in the reality is closely connected with the kinetic features of the polymer formation way.

SOME KINETIC PECULIARITIES OF POLYCONDENSATION AND POLYMERIZATION

Polycondensation process can be described in a common case as three stage reaction:

$$
\begin{aligned}
M_{1,k} + M_{1,k} &\longrightarrow R_{2,2(k-1)} & k_1 \\
R_{i,j} + M_{1,k} &\longrightarrow R_{i+1,j+k-2} & k_2 \\
R_{i,j} + R_{m,n} &\longrightarrow R_{i+m,j+n-2} & k_3
\end{aligned}
\tag{3}
$$

characterized by different rate constants k_1, k_2 and k_3 of each stage. Here, first index denotes, as in the equation (1), a number of monomer units and the second one a number of functional groups in a chain.

For the description of the conversion two different definition can be used: conversion of monomer

$$
\alpha_m = i - M_1 / M_0
\tag{4}
$$

and conversion of functional groups

$$
\alpha_f = 1 - \sum_i f_i \cdot R_i \ / f \cdot M_{1,0},
\tag{5}
$$

where $f_i = 2 + (f-2) \cdot i$ and $f \geq 2$.

Let us assume that $k_2 / k_1 = k_3 / k_2 = \gamma$. In this case the correlation between α_m and α_f will be as follows:

$$
\alpha_m = \begin{cases}
\alpha_f & \text{if } \gamma >> 1 \\
1 - (1 - \alpha_f)^f, & \text{if } \gamma = 1 \\
f \cdot \alpha_f & \text{if } \gamma << 1
\end{cases}
\tag{6}
$$

To compare these results with those for polymerization we have to consider the following kinetic scheme:

$$
R_{i,j,k} + M_f \longrightarrow R_{i+1,j+f-1,k} \qquad k_p
\tag{7}
$$

$$R_{i,j,k} + R_{m,n,l} \xrightarrow{\quad k_c \quad} Ri_{+m,j+n-1,k+l}$$

where k_p and k_c are the rate constants of chain propagation and crosslinking respectively, M_f is f-functional monomer, $f \geq 1$, while the other symbols are the same as in the scheme (2). If we define α_m and α_f in the following way:

$$\alpha_m = 1 - M / M_0 \tag{8}$$

and

$$\alpha_f = 1 - (\sum j \cdot R_{i,j,k} + f \cdot M) / f \cdot M_0 \tag{9}$$

and assume that $k_c/k_p = \gamma$, the correlation between α_m and α_f turn out to be the same as it was shown in the equation (6). It is not surprising, because in both cases the value γ characterizes degree of the substitution effect, i.e. relative rate of the consumption of a functional group and monomer. So, the relationship between consumption of a monomer and functional groups depends on the substitution effect only and does not depend on the way of polymer formation.

Nevertheless, great differences of polycondensation and polymerization can be observed in the regularities of the polymer degree of polymerization growth as a function of conversion. Such dependencies for the f-functional polycondensation (eq. 10) and f-functional "living" ionic and radical polymerization (eqs. 11 and 12) are presented below.

$$\bar{P}_N = 1/(1 - f \cdot \alpha_f /2) \tag{10}$$

$$\frac{\bar{P}_N}{\bar{P}_{N,0}} = [1 - \gamma \cdot \frac{f-1}{2 \cdot f} \cdot \bar{P}_{N,0} \cdot \alpha_m]^{-1} \tag{11}$$

$$\frac{\bar{P}_N}{\bar{P}_{N,0}} = [1 - \frac{1}{2} \cdot \gamma \cdot (f-1) \cdot \frac{k_p \cdot M_0}{k_t \cdot R} \cdot \frac{2 \cdot \alpha_m}{2 + \alpha_m}]^{-1} \tag{12}$$

In these equations $\bar{P}_N$ is a number average degree of polymerization, $\bar{P}_{N,0}$ is $\bar{P}_N$ at $\gamma = k_c/k_p = 0$, R is a steady state radical concentration ($R = \sum k \cdot R_{i,j,k}$) and k_t is the rate constant of termination reaction at radical polymerization:

$$R_{i,j,k} + R_{m,n,l} \xrightarrow{\quad k_t \quad} R_{i+m,j+n,k+l-2} \tag{13}$$

Other symbols used in eqs. (10-12) are the same as used before.

As one can see the eqs. (10-12) are quite different. At polycondensation the $\bar{P}_N$ growth rate is very slow up to the gel-point. Quite a different situation takes place at polymerization. As it is seen from eqs. (11) and (12) the second terms of these equations stipulate rapid growth of $\bar{P}_N$ with the growth of α_m. The dependence of $\bar{P}_N(\alpha)$ can be very visibly demonstrated for the well-studied processes of linear polymers formation (Fig. 1).

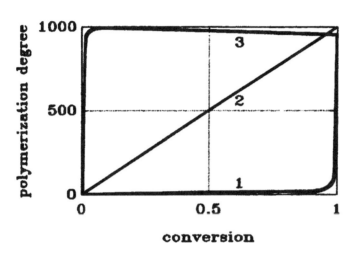

Figure. 1. Dependencies of number average degree of polymerization on conversion of linear polymer for polycondensation (1), "living" ionic polymerization (2) and radical polymerization (3).

It is worth noting that for obtaining high molecular weight polymer, as it is seen from Fig. 1, the "living" ionic polymerization is much more preferable in comparison with polycondensation. By the way, a great variety of heterochain polymers, that had been prepared before by polycondensation only, now can be also prepared by ionic polymerization owing to the invention of General Electric chemists [11]. It is very difficult to obtain high molecular weight polymer by polycondensation even if there are no side reactions and the stoichiometry of reacting functional groups are preserved in the course of the reaction. It is connected with the specific dependence of $\bar{P}_N$ on α: it is rather weak up to the very high conversion and $\bar{P}_N$ begins to grow rapidly only in the very last reaction stages. However it takes a lot of time and an essential reaction time is wasted on this stage. Fig. 2 obviously demonstrates this conclusion. On this figure Δt is time that is necessary for the increasing of $\bar{P}_N$ in x times ($x = \bar{P}_{N,t} / \bar{P}_{N,to}$).

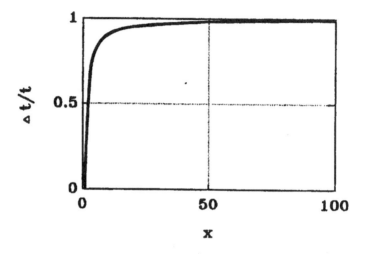

Figure 2. Dependence $\Delta t / t$ on x.

There is no need to try to reach very high conversion (0.999) at "living" ionic polymerization for obtaining high molecular weight polymer, because in that case molecular weight linearly increases with the conversion and can be controlled by a catalyst concentration.

Dependencies of $\bar{P}_N(\alpha)$ naturally are not so important for network polymers as for linear ones. Nevertheless, they determine such critical phenomenon as gelation, i.e. attaining the conversion when $\bar{P}_W \longrightarrow \infty$. For qualitative comparison of different ways of polymer formation we can choose also the critical condition $\bar{P}_N \longrightarrow \infty$. It will be an upper estimation. As it follows from eqs. (10-12) at equal monomer functionality the gel-point occurs at much more high conversion at polycondensation then at radical polymerization. In the latter case it is not far from zero conversion at $f \geq 2$. Gel-point for the "living" ionic polymerization is among two cases mentioned above. The functionality of forming polymer chains increases far more rapidly at polymerization than at polycondensation. It is the result of the fact that at the same conversion of functional groups the number of polymer chains at polycondensation is functional groups the number of polymer chains at polycondensation is considerably higher than at polymerization, when the number of polymer chains depends on the rate of initiation. This reason stipulates the local proceeding of polymerization in comparison with polycondensation. The reaction occurs with the highest probability in places where highest molecular weight chains are localized. For the radical polymerization, because of radical life-time is much less than the whole time of polymerization, an ability to local proceeding of a reaction is far more pronounced.

INTRAMOLECULAR REACTIONS OF CYCLIZATION

Both polycondensation and polymerization reactions are usually accompanied by intramolecular reactions of cyclization. Such reactions in the case of linear chains can proceed by two mechanisms:

1. End-to-end cyclization, i.e. the reaction of two terminal functional groups at polycondensation or active propagating center with the terminal functional group at polymerization.

2. So called back-biting reaction, when an active propagating center or terminal functional group attacks randomly any link of their own chains, forming macro-cycle and resuming an active center or terminal functional group.

The second type of intramolecular reactions usually is the most characteristic for the cationic polymerization of heterocycles and polycondensation reactions, leading to the formation of heterochain polymers.

One of the main differencies between polycondensation and polymerization consists in difference of dependence $\bar{P}_N = f(\alpha)$. This difference is exhibited also in the character of accumulation of macrocycles. Below we will examine the results of such analysis for these two types of intramolecular reactions. Here we consider such intramolecular reactions in which the driving force for the cyclization is an entropy change only and kinetics of macrocycles formation can be described in the term of Jacobson-Stockmayer theory [12].

1. Kinetics of cycles accumulation for first mechanism can be described by the equation (14):

$$\frac{dQ_i}{dt} = k \cdot i^{-3/2} \cdot R_i , \qquad\qquad (14)$$

2. Kinetic equation describing the rate of accumulation of the cyclic oligomers during the back-biting reaction has the following view,

$$\frac{dQ_i}{dt} = k \cdot i^{-3/2} \cdot \sum_{j=i}^{\infty} j \cdot R_j \qquad (15)$$

where Q_i and R_i are the molar concentrations of cycles and linear chains, containing i monomer units, respectively, k is the rate constant of cyclization reaction.

By equations (14) and (15) we can calculate the rate of change of the first three moments of molecular weight distribution for cyclic molecules formed by end-to-end and back-biting cyclization. So, we will analyze the rate of accumulation of mass concentration of cyclic oligomers

$$m'_c = (\sum i \cdot Q_i) / dt \qquad (16)$$

and instantaneous weight average polymerization degree of cyclic oligomers $(\bar{P}'_{w,c})$

$$\bar{P}'_{W,c} \equiv \frac{d \ i \cdot Q_i^2}{d \ i \cdot Q_i} \qquad (17)$$

for polycondensation and some cases of polymerization.

Let us assume that the distribution of macromolecules of sizes is described by the Flory-Schulz type:

$$R_i = A \cdot i^{\alpha} \cdot e^{-\beta \cdot i} \qquad (18)$$

where parameters α and β of this distribution equal:

$$\alpha = \frac{2 - \bar{P}_W / \bar{P}_N}{\bar{P}_W / \bar{P}_N - 1} \qquad and \qquad \beta = (\alpha + 1) / P_N \qquad (19)$$

A is normalizing factor, $\bar{P}_W$ and $\bar{P}_N$ is weight and number average polymerization degree of linear polymer respectively.

Dependencies between the concentrations of cyclic (m'_c) and linear **(m)** molecules and $\bar{P}'_{w,c}$, and $\bar{P}_W$ obtained from equations (14-19) are given below.

	end-to-end cyclization	back-biting cyclization
m'_c	$A \cdot k \cdot \bar{P}_N^{-3/2} \cdot m$	$A \cdot k \cdot \bar{P}_N^{-1/2} \cdot m$ (20)
$\bar{P}'_{W,c}$	$B \cdot \bar{P}_W$	$B \cdot \bar{P}_W$ (21)

Numerical coefficients **A** and **B** depending on the way of polymer formation are summarized in the Table 1.

Table 1. Numerical coefficients in the equations (20-21).

Type of reaction	$\dfrac{\bar{P}_W}{\bar{P}_N}$	Type of cyclization			
		"end-to-end"		"back-biting"	
		A	B	A	B
Polycondensation	2	1.77	0.25	2.65	0.42
Radical polymerization	1,5	1,25	0,5	2,34	0,39
"Living" ionic polymerization	1	1	1	2	0.33

As it is seen from the equations (20) and (21) and Table the rate of cyclic oligomers accumulation is proportional to the mass concentration of the linear polymer formed for both cyclization mechanisms, inversely proportional to the $\bar{P}_N^{3/2}$ for the "end-to-end" cyclization and straight proportional to the $\bar{P}_N^{1/2}$ for the "back-biting" one. The rate of cyclic molecules accumulation is determined by the character of the polydispersity of polymer formed and it is slightly different for polycondensation, radical "living" polymerization.

The instantaneous weight average polymerization degree of cyclic oligomers is straight proportional to the weight average polymerization degree of the linear polymer formed for both mechanisms of cycles formation. It does not practically depend on the mechanism of linear chains formation in the case of "back-biting" cyclization, while in the case of "end-to-end" cyclization $\bar{P}'_{W,c}$ increases four times with the decreasing of polydispersity coefficient from 2 to 1.

So, the kinetic laws of cyclization during linear polymer formation are the same with the accuracy of the numerical coefficients for all types of polymer formation processes considered above. They vary depending on the cyclization mechanism only. It means that the main factor, which affects the fraction of cyclic molecules, formed during polymer formation is the character of the polymerization degree growth with conversion.

Cyclization reactions considered above for the processes of network formation are the reactions of ineffective cyclization. They affect the value of gel-point, sol-fraction, phase transformation in the reacting system and, finally, the topological structure and morphology of polymer formed. From the kinetic point of view the cyclization reactions are a termination reaction of the network chain development [1]. Their competition with the crosslink reaction determines the results of the formation of the macromolecule with infinite size i.e. gelation of the system. This is analog to chain explosion under the proceeding of the branched chain reaction according to Semenov [13]. Unfortunately, nowadays it is impossible to describe this situation quantitatively but some qualitative examples are considered in [1].

TOPOLOGICAL STRUCTURE OF NETWORKS

Topological structure of network polymers obtained by polycondensation and polymerization reflects the kinetic peculiarities of their formation, as well as other levels of structural organization of polymer systems. It is generally assumed, that topological structure is one of the levels of the polymer system organization that characterizes the connectivity of the elements of the system and can be expressed as a graph without considering the concrete chemical structure and arrangement of their elements in space [1].

What will be the topological structure for the ideal network polymers with $\alpha_f = 1$, obtained by f-functional monomers polycondensation and polymerization? It is obvious, they will be quite different. The topological structure of polycondensation network in the case under consideration can be characterized by different types of cycles only without any dangling chains. In the case of network, obtained by polymerization, topological structure of the polymer will be characterized at least by two types of elements: cycles and dangling chains. In reality these differences of networks increase only since it is practically impossible to attain not only $\alpha_f = 1$, but also $\alpha_m = 1$ in the case of polymerization because of the diffussional control of the chain propagation on the deep reaction stages [1,3,7]. The second reason of increase of the number of tails in the network is any kind of transfer reactions, increasing the number of chains in the system and, consequently, the number of dead terminal groups. It is worth noting that in the case of a network, formed by polycondensation equality $\alpha_{m,\infty} = 1$ is always true, but $\alpha_{f,\infty} < 1$ (where $\alpha_{m,\infty}$ and $\alpha_{f,\infty}$ denote limit values of the corresponding conversions at $t \longrightarrow \infty$). This limitation of $\alpha_{f,\infty}$ under the conditions, when the composition of reacting functional groups is close to stoichiometric one, and cure temperature is higher than maximal value of Tg for the given network polymer, is related to the topological limit of the cure process [10,14,15]. In this case the translational mobility of the functional groups in the system with very complex topological structure at high conversion ($\alpha_{f,\infty} \longrightarrow 1$) is close to 0 in the rubbery state and cure process is practically stopped.

Pulse NMR is the most effective and simple modern method for the observation of the forming network polymer topological structure evolution. This approach is based on the fact that network elements of various topological complexity (branching units with different functionality) posses different relaxation properties. The main advantage of the method consists in the possibility of the continuous monitoring of the system during cure and the elucidation of its kinetic inhomogeneity. It can be done by the evaluation of relaxation times spectrum width, as well as the determination of the proton fraction characteristic to each kinetic phases (components of the system with different mobility), when analyzing the curve of free induction decay [16-19].

On the other side, the proton fraction of the different type of hypothetical kinetic units (branching units with different functionality, dangling chains and chains between crosslinks) can be calculated from statistic theory [1,3-6,20]. The comparison of these calculated values of proton fraction with the measured ones helps to understand the mechanism of the relaxation process itself. This approach allows one to make some conclusions about the features of the developing topological structure of the networks obtained by polycondensation and polymerization. Such experimental investigations by pulse NMR method [17-19] based on the approach considered above completely confirm the qualitative conclusions about the features of the topological structure of the network polymers obtained via polycondensation and polymerization.

PHASE SEPARATION DURING NETWORKS FORMATION

Reactions of cyclization are an intrinsic part of network formation processes. They lead to the gelation that takes place at the very low conversion for polymerization and proceeds locally accompanied by the microgel formation. The reactions of microgel formation and following reactions of functional groups inside microgel have an obviously expressed autoaccelerated character [1] because of the network elasticity manifestation. It leads to the contraction of microgel and as a consequence pushing out from microgel all low-molecular compounds, and finally to the phase separation in the system. This phenomenon has been studied by Dusek [21,22] in detail and now is known as microsyneresis (v- and χ-microsyneresis). The phenomenon of microsyneresis during the network formation is expressed especially brightly for the radical polymerization, in which from the very beginning of process the diluted solution of polymer in own monomer is formed. Such situation is very favorable for the local proceeding of intramolecular crosslinking whose probability in such conditions is close to 1. Such reactions result in sharp contraction of microgel and finally the new phase formation. The following microgel propagation looks like spreading of reaction front. Then different microgels come into contact with each other and finally all system acquires the monolithic character with the network defects between microgels (or grains in glassy state)[*].

So, the phenomenon of microsyneresis along with gel-effect is the powerful factor of autoacceleration and one of the important reasons for inhomogeneous development of polymerization, and as a consequence for the microheterogeneous character of forming network polymer.

The processes of network polymer formation by polycondensation usually proceed according to the really statistical laws and homogeneous polymers are formed [1,20]. Microsyneresis can take place but at sufficiently high conversion, when the initial monomers in the system practically completely exhaust. Therefore microsyneresis in polycondensation is exhibited far less brightly than

---

[*]   In the case under consideration the local development of the reaction is caused also by such kinetic factor as "gel-effect" [1,3-5,7].

in polymerization. It is worth underlining that this conclusion is related to both n- and c-microsyneresis. However, at the presence of solvent, plastisizer or other additives i.e. in multicomponent systems, microsyneresis frequently occurs and additives usually become a dispersed phase. The main driving forces in these cases are not the network formation (n-microsyneresis) as in polymerization [23-26].

Continuous change of the thermodynamic parameters during the process of network formation results in the change of the new phase nucleation probability [24-26].

In the same time continuous change of the diffusion coefficient during the network formation results in a change of the new phase particles growth rate. Because of that the volume fraction of dispersed phase and phase structure are determined largely by kinetics of polymer formation [24-26]. If the nucleation of new phase takes place at high value of the diffusion coefficient, i.e. at low monomer conversion, the growth of supersaturation in the system is compensated by the diffusion stream of substance to the growing particle. In this case the nucleation is suppressed by the particles' growth. If the diffusion coefficient becomes very low (high conversion) the particles' growth is suppressed and intensive nucleation occurs. Under certain conditions it can result in the appearance of the second or even third mode on the distribution curve of dispersed phase particles for their size [23-25].

So, the phase structure of modified heterogeneous networks obtained under the conditions of phase separation of modifier during the polymerization or polycondensation reaction is determined by the competition of two kinetic factors: rate of chemical reaction, which determines the dependence of the equilibrium concentration of a modifier on conversion, and the diffusion coefficient, determining the current concentration of modifier in the solution, and finally the supersaturation in the system. As far as the kinetic laws of polycondensation and polymerization are quite different, therefore the regularities of phase separation for these two types of processes of a polymer formation as well as the phase structure of resulting polymer (even of the same chemical structure) have to be quite different.

INHOMOGENEITY OF NETWORK POLYMERS

The analysis made above shows that one of the main features of polymerization in comparison with polycondensation is its tendency to inhomogeneous development of the process, and as a result the formation of polymer with inhomogeneous topological and supermolecular structure. Here we will try to formulate the conditions for inhomogeneous proceeding of the process of polymer formation, and, as a consequence formation of the inhomogeneous polymer.

The reason for inhomogeneous development of the reaction can be revealed by analysis even of the simplest model system: monomolecular autocatalytic reaction in

$$A \xrightarrow{k_1} B; \qquad\qquad A + B \xrightarrow{k_2} 2B \qquad (22)$$

conditions of limited mobility of the reagents [27]. The analysis was carried out by computer simulation using Monte-Carlo method. As was expected statistically uniform distribution of the reaction product molecules in the space is observed in the absence of autocatalysis ($k_2 = 0$) irrespective of the reagents' mobility in the system. The increasing of catalytic rate constant k or reagents mobility in the reaction system leads to the growth of the reaction products clusterization degree. High mobility of reagents suppresses in this case the action of autocatalytic reaction on the distribution of the reaction products in the reaction system (Figures 3-5).

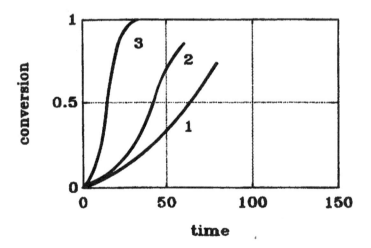

Figure 3. Kinetic curves of the monomolecular autocatalytic reaction with limited mobility of reagents. Reagents' mobility in the system p equal to number

of movement acts per one chemical reaction act: 1 - 0; 2 - 10; 3 - 100. Factor of autocatalyticity $r = k_2 \cdot A_0/k_1 = 70$.

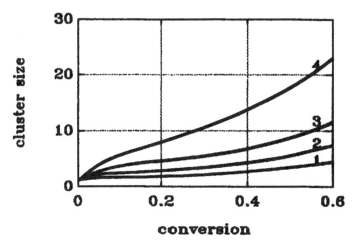

Figure 4. Change of the number average cluster size of reaction products with conversion for the monomolecular autocatalytic reaction. p: 1-100; 2-10; 3-1; 4 - O. r = 70.

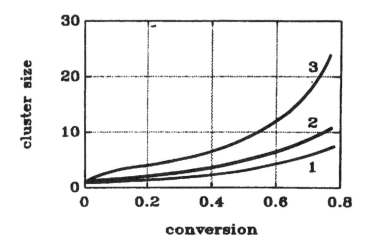

Figure 5. Dependence of the number average cluster size of reaction products on conversion for monomolecular autocatalytic reaction. r: 1 - 70; 2 - 1,7; 3 - 0.7. p=1.

As was expected the limit conversion in the case under consideration does not depend on the mobility of reagents in the system.

Similar results have been obtained for bimolecular autocatalytic reaction:

$$A + B \xrightarrow{\ k_1\ } 2C; \qquad\qquad A + B + C \xrightarrow{\ k_2\ } 3C \qquad\qquad (23)$$

It is worth noting that in contrast to the previous case (scheme 22), in the absence of reagents mobility in the system the reaction stops after the conversion become equal to 0.625 independently of degree of autocatalysis. This result is quite clear: the reaction limit is typically topological by its nature and is determined by the character of arrangement of components in the space in initial state that was random in the case under consideration.

What conclusion for polymer system we can make from these results? In the case of polymer formation the continuous growth of the reaction system viscosity is favorable for inhomogeneous development of the reaction, but it can be realized in practice only if the polymer formation reaction develops with autoacceleration. This is the obligatory condition. The mechanism of autoacceleration can be of character either chemical (autocatalysis) or physical (gel-effect in radical polymerization, microsyneresis, etc.). It is worth noting that even if the process of a network polymer formation occurs without any autoaceleration, the kinetic curve of weight average functionality of growing molecules has autoaccelerated form. Autoacceleration of a chain growth is inherent feature of network formation processes because they are developed as branched chain reaction. The rate of the molecules' functionality growth is much faster for polymerization than for polycondensation.

So, from the very common consideration the process of the network polymer formation via polymerization in comparison with polycondensation has greater probability of localization in space, and therefore of formation of the networks with inhomogeneous topological and supermolecular structures.

CONCLUSIONS

The main peculiarity that differentiates the processes of polymer formation one from another is the character of polymer chains growth. Namely this feature determines the tendency to differences in topological structure organization of network polymers obtained by polymerization and polycondensation even if their molecular structure is the same.

This peculiarity along with some specific thermodynamic and kinetic effects (microsyneresis, gel-effect, autocatalysis, etc.), stipulates the tendency to the localized development of polymerization in comparison with polycondensation and to formation of polymers with inhomogeneous or even phase separated supermolecular structure.

REFERENCES

[1] V.I.Irzhak, B.A.Rozenberg, N.S.Enicolopyan, Network Polymers: Synthesis, Structure, Properties, (in Russian), Science, M., (1979).

[2] Al.Al.Berlin, S.A.Volfson, N.S.Enicolopyan, Kinetics of Polymerization Processes, Chemistry, M., (1978).

[3] P.C.Hiemenz, Polymer Chemistry. The Basic Concepts, M.Dekker Inc., N.Y., Basel, (1984).

[4] Kuchanov S.I., The Methods of Kinetic Calculations in Polymer Chemistry, (in Russian), Chemistry, M., (1978).

[5] C.Tanford, Physical Chemistry of Macromolecules, Wiley, N.Y. (1961).

[6] S.I.Kuchanov, S.V.Korolev, S.V.Panyukov, Adv. Chem. Phys., 72, 115-326, (1988).

[7] A.A.Berlin, G.V.Korolev, T.Ya.Kefeli, Yu.M.Sivergin, Acrylic oligomers and materials on their basis, (in Russian), Chemistry, M., (1983).

[8] K.Dusek, J.Plestil, F.Lednicky, S.Lunak, Polymer, 19, 373 - 397, (1978).

[9] R.J.Morgan, Adv. Pol. Sci., 72, 1, (1985).

[10] V.I.Irzhak, B.A.Rozenberg, Vysokomol. soed., 27A, 1795-1808, (1985).

[11] D.J.Brunelle, E.P.Boden, T.G.Shannon, J. Am. Chem. Soc, 112, 2399-2402, (1990).

[12] H.Jacobson, W.H.Stockmayer, J.Chem.Phys., 18, 1600 - 1612, (1950).

[13] N.N.Semenov, Chain reactions, L., (1934).

[14] V.A.Topolkaraev, Ph.D.Thesis, Inst. Chem. Phys. USSR Academy of Sciences, M., (1977).

[15] E.F.Oleinik, Adv. Pol. Sci., 80, 49 - 99, (1986).

[16] A.I.Maklakov, V.S.Derinovskii, Usp. Khim., 48, 749 - 771, (1979).

[17] V.I.Irzhak, V.M.Lantsov, B.A.Rozenberg, in: "Cross linked Epoxies",

pp.359-371. W.de Gruyter Co., Berlin-N.Y. (1987).

[18] I.N.Zakirov, V.I.Irzhak, V.M.Lantsov, B.A.Rozenberg, <u>Vysokomol.soed.</u>, <u>30A</u>, 915-921, (1988).

[19] V.I.Lantsov, Dr.Sc. <u>Thesis, Inst. Chem. Phys.</u> USSR Acad.Sci., M., (1989).

[20] K.Dusek, <u>Makromol. Chem.</u>, <u>Macromol. Symp.</u>, 7, 37 - 53 (1987).

[21].K.Dusek, <u>J. Pol. Sci.</u>, <u>C 16</u>, 1289 - 1299, (1967).

[22] K.Dusek, <u>J. Pol. Sci.</u>, <u>C 39</u>, 83 - 106, (1972).

[23] A.Vazquez, A.j.Rojas, H.E.Adabbo, J.Borrajo, R.J.J.Williams, <u>Polymer</u>, <u>28</u>, 1156 - 1164, (1987).

[24] G.F.Roginskaya, V.P.Volkov, E.A.Dzhavadyan, G.S.Zaspinok, B.A.Rozenberg, N.S.Enikolopyan, <u>Dokl. Akad. Nauk SSSR</u>, <u>290</u>, 630 - 634, (1986).

[25] B.A.Rozenberg, <u>Problems of phase formation in oligomeric systems</u>, (in Russian), Chernogolovka (1986).

[26] B.A.Rozenberg, <u>Makromol. Chem.</u>, <u>Macromol. Symp.</u>, <u>41</u>, 165-177, (1991)p

[27] V.I.Irzhak, N.I.Peregudov, B.A.Rozenberg, N.S.Enikolopyan, <u>Dokl. Akad. Nauk SSSR</u>, <u>263</u>, 630-633, (1982).

Polymer Networks '91 pp. 25-38
Dosek and Kuchanov (Eds)
© VSP 1992

Network formation via end-linking processes

Paul Rempp, René Muller and Yves Gnanou

Institut Charles Sadron CNRS-ULP, 6, rue Boussingault, 67083 Strasbourg Cedex, France

ABSTRACT

Polyurethane networks exhibiting hydrophilic elastic chains have been synthesized by stoichiometric end-linking reactions. The precursor polymers were either poly(ethylene oxide) [PEO] or polydioxolane [PDXL], of known molecular weight, fitted at both chain ends with OH functions. Various pluriisocyanates [Desmodur] served as the antagonist reagents in these step growth crosslinking reactions.

An investigation of the crosslinking process was performed by following the rheological behavior of the reaction medium as end-linking proceeds. The storage G' and loss G" moduli were measured at various stages of the reaction, over a wide range of oscillatory shear frequencies. Simultaneously, the conversion attained was followed by Fourier Transform Infrared. These kinetic and rheological data were accounted for on the basis of the Miller-Macosko simulation based on conditional probabilities. With stoichiometric systems, the agreement is very good between the values of the gel point found and those calculated. It is confirmed that at the gel point, the storage and loss moduli are congruent, regardless of frequency, and that the variation of G' and G" with ω is that observed by Winter and Chambon. This investigation was extended to the case of non-stoichiometric reaction mixtures, especially those involving an excess of precursor.

Networks can be synthesized in various manners : by crosslinking polycondensations, by free radical copolymerizations involving a bis-unsaturated comonomer, by random couplings (bridging) between existing polymer chains and by end-linking reactions. The latter method actually exhibits many similarities with crosslinking polycondensations, the main difference is that the bifunctional constituent is itself polymeric. Control of the

length and of the number of elastic chains is thus provided. The
basic features of end-linking processes are as follows[1] :

A linear, well-characterized, α,ω-difunctional precursor
polymer (A$\sim\!\sim\!\sim$A) is reacted stoichiometrically with a plurifunctional low molecular weight compound B_f. Reaction of function A
with B builds a link. At a given conversion the gel point is
attained, but the reaction is pursued, in order to allow each A
function to find a B partner to react with. At quantitative
conversion, each precursor molecule has become an elastic chain,
and each B_f molecule has become a branch point, connected with as
many chain ends as there were functions initially : f. These end-
linking reactions are generally carried out in the presence of a
solvent, in order to confer enough mobility to the chains and
accessibility to the functions, so as to allow high conversions to
be attained. The concentration has to be chosen high enough,
however, to keep the chances for loop formation as low as possible.

Stoichiometric end-linking reactions carried out to complete
conversion are expected to yield well defined polymer networks,
i.e. species constituted of a known number ν of elastically
effective chains of known length, connected to μ branch points,
distributed quite homogeneously[2]. It can be anticipated, however,
that such networks contain a few unavoidable defects :

* <u>dangling chains</u> can arise either from incomplete conversion or
from a local excess of A functions. [If the stoichiometric ratio
r - [B]/[A] is lower than unity, then the existence of dangling
chains is quite obvious].
* <u>loops</u> are formed upon reaction of both ends of the same A$\sim\!\sim\!\sim$A
precursor molecule with two B functions located on the same B_f
molecule.
* <u>double connections</u> imply that two different chains connect the
same two branch points.

In a stoichiometric end-linking reaction, as in any
crosslinking polycondensation, the number average molecular weight

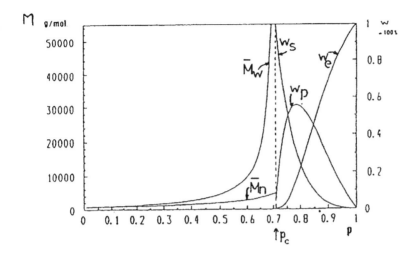

Fig. 1 - Schematic representation of the variation of molecular weight averages and of the weight fractions of extractable chains, of elastically effective network chains, and of dangling chains versus conversion, in an end-linking reaction.

increases slowly as conversion progresses, but the weight average molecular weight rises faster and it eventually reaches infinity at the gel point (figure 1). At this stage most of the molecules present in the reaction medium are still of finite size, and they can be extracted from the gel[3]. As conversion progresses beyond the gel point, the amount of soluble material still present in the network decreases and it tends towards zero as conversion becomes quantitative.

It was found earlier that simulations following the theory of Miller and Macosko[4,5], based on conditional probabilities of reaction, are in good agreement with the experimental results : the amount extractable material found in the networks at given conversions was very close to that predicted by the simulation. As cyclizations are not considered in the Miller-Macosko model, it was concluded that under the experimental conditions chosen, the chances for loop formation are negligible.

The structural homogeneity of these networks has been questioned, however, on the basis of inelastic light scattering results and of X-ray and neutron scattering experiments[6]. In networks swollen to equilibrium, rather large fluctuations in segment densities - and in crosslink densities - have been detected within a sample. A picture featuring clusters - exhibiting large segment densities - loosely connected to each other has been proposed to account for the experimental observations. This model is still somewhat controversial, and this aspect will not be discussed further here. Despite their presumed "inhomogeneities", networks arising fromd end-linking processes are of great interest because the average length of the elastic chains is known (from the molecular weight of the bifunctional A$\sim\sim$A precursor polymer). Similarly, the functionality of the branch points arises from that of the reagent B_f, if quantitative conversion have been attained and if no side reactions have occurred. End-linking reactions usually proceed smoothly, in homogeneous phase, and the method is applicable to a large number of systems[7].

An interesting method to understand the essential features of end-linking reactions, is to follow the rheological behavior of the reaction medium as conversion progresses. The rheological response of any viscous solution or swollen gel is known to be frequency dependent. It has been first established by Tung and Dynes[8] and by Adam, Delsanti and Durand[9,10] that the transition from viscous liquid to solid gel -the gel point - can be determined precisely on the basis of rheological measurements. Winter and Chambon[11,12,13] have shown that the storage and the loss moduli are congruent at the gel point, over a wide range of oscillatory shear frequencies applied.

Our contribution[14] to the investigation of network formation by end-linking processes involved simultaneous determinations of the conversion attained and of the storage and loss moduli of the reaction medium (over a wide range of oscillatory frequencies), as reaction proceeded. Special attention was devoted to the phenomena occuring in close vicinity of the gel point.

EXPERIMENTAL INVESTIGATION OF THE GELATION PROCESS

We have studied the end-linking reaction between α,ω-dihydroxy-poly(ethylene oxide) precursor polymers and plurifunctional isocyanantes[2]. Other experiments were run with α,ω-dihydroxy-poly-dioxolane as the precursor polymer[15,16]. The precursors were accurately characterized and their molecular weight distributions were shown to be quite narrow. The polyurethane networks arising from these reactions are of interest, as they are constituted of hydrophilic (PEO or PDXL) network chains, connected by means of hydrophobic branch points.

The end-linking reactions were performed at a fixed concentration of 33 wt-%, in pure dioxane, in the absence of catalyst, and at a temperature (60°C) chosen to minimize side reactions[2], such as allophanate formation. Three different methods were considered to investigate the network forming process.

1 - **Fourier Transform Infrared** (FTIR) was used to determine the conversion attained as a function of reaction time. The FTIR cell containing the reaction mixture was kept at constant temperature. Both the consumption of isocyanate functions and the formation of isocyanate linkages were measured accurately, at fixed time intervals, allowing thus a double-check of the number of links formed between precursor chain ends and branch points as reaction proceed.

2 - **The rheological behavior of the reaction mixture** was studied by means of a mechanical spectrometer Rheometrix, at various stages of the end-linking reaction. The cone-and-plane cell was filled with the same reaction mixture as the FTIR cell, and kept at the same temperature. At fixed time intervals, the loss and storage moduli of the reaction medium were measured over a wide range of oscillatory shear frequencies.

3 - **The Miller-Macosko simulation** was applied to calculate, for each reaction mixture, the conversion at which gelation should be expected to occur. The method was also used to determine, for any

conversion beyond the gel point, the amount of extractable material
to be expected. The parameters required for this calculation are
the functionality f of the reagent B_f and the stoichiometric   ratio
r - [NCO]/[OH].   Cyclizations   are   ignored, thus concentrations
should not be not chosen too low.

EXPERIMENTAL RESULTS

A - Stoichiometric reaction. mixtures : r - 1

If the storage modulus G' and the loss modulus G" are
plotted versus reaction time for a given frequency of oscillatory
shear applied (figure 2), the   two curves cross each other   after a
given reaction time t_c, at which gelation occurs. t_c is the same
regardless of the frequency of the shear applied.

Storage and loss moduli can also be plotted (in logarithmic
scale) as a function of the frequency of the oscillatory shear
applied, for given values of the reaction time. At t_c it is
confirmed that G' and G" overlap over the entire frequency range
(Figure 3). The slope of this common line is 1/2, which means that,
at the gel point, G' and G" are proportional to $\omega^{1/2}$. At reaction
times shorter than t_c, G" is higher than G', and at times past t_c
the reverse is true.

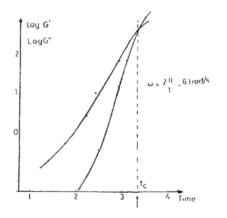

Fig. 2 - Plot of the log of storage and loss moduli, as a function
of reaction time, at constant frequency of the shear applied.

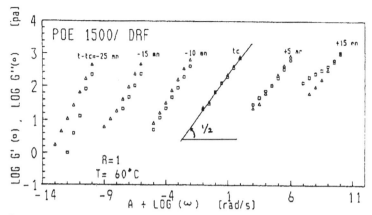

Fig. 3 - Variation of the storage and loss moduli G' and G" versus frequency of the shear applied, in log scale, at various reaction times. t_c stands for the time at which gelation occurs.

From the FTIR data obtained on the same reaction mixture, under the same conditions, a relation is established between reaction time t and the conversion attained p. Thus, plots of log G' and of log G" versus p can be set up. The diagram shown on figure 4 refers to one single frequency but the G' / G" crossover occurs at the same value of p, regardless of the frequency. p_c is found to be equal to 0.71, in excellent agreement with the value predicted by the Miller-Macosko theory for that system (0.707).

Figure 5 shows another example, in which the average functionality of the (polydisperse) crosslinking agent is far higher than in the previous case. Again log G' and log G" have been plotted versus log ω, for given reaction times t. At t_c (corresponding to a conversion p_c) G' and G" overlap in the entire frequency range studied, and the slope of the common line indicates a linear variation of G' and of G" with $\omega^{1/2}$. Gelation occurs here at a conversion p_c - 0.32. The corresponding value calculated according to the Miller-Macosko theory is 0.33.

To get a more accurate insight into the rheological behavior of the reaction medium shortly before gelation, an experiment was run in which the reaction was stopped at $t - t_c - 15$ minutes, by addition of a small amount of butyl amine. The viscous solution was

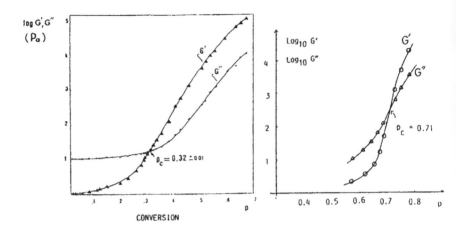

Fig. 4 - Variation of the storage and loss moduli, in log scale, as a function of the conversion attained, at fixed shear frequency.

Fig. 5 - Same plot as Fig 4, for a stoichiometric system involving a urethane, with a number average functionality of 6.5.

then submitted to oscillatory sollicitations over a wide range of frequencies. In the low frequency range G' is proportional to ω^2 and G'' is proportional to ω. In the high frequency range the two curves tend to become parallel and to vary linearly with $\omega^{1/2}$. Since the reaction has been stopped prior to gelation, it is quite obvious that G'' is still above G' (Figure 6).

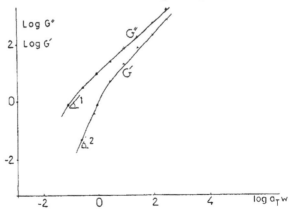

Fig. 6 - Master curve of the storage and loss moduli plotted in log scale versus $a_T\omega$ for a stoichiometric system, the reaction having been stopped 15 minutes before gelation.

This behavior reminds of that exhibited by concentrated solutions of linear polymers. Two regions have to be distinguished in the plots of G' and G'' (in log scale) versus ω. At low frequencies, the slopes are respectively 2 and 1, whereas at high frequencies both moduli exhibit a common exponent, close to 1/2. The boundary between these two regions is shifted towards lower frequencies as the molecular weight increases. Similar observations have been made by Winter and Chambon[12].

In the case of end-linking processes, this crossover frequency should eventually reach zero as the reaction reaches gelation. Beyond the gel point, Rouse-like behaviour thus extends throughout the range of frequencies. The reaction medium then behaves neither as a liquid (implying that $G' \approx \omega^2$ and $G'' \approx \omega$) nor like an elastomeric solid (in which G' is a constant, $G\infty$, while G'' is still proportional to ω). This is precisely what has been observed : Beyond the gel point G' tends to level off, whereas G'' is still increasing as the oscillatory frequency applied increases.

Thus in a stoichiometric reaction mixture the observed rheological response of the medium to oscillatory shear covering a wide range of frequencies does not deviate significantly from the predictions of the Rouse model, once gelation has occurred.

B - Non-stoichiometric reaction mixtures : $r \neq 1$

1 - If the B functions are in excess. ($r > 1$) the network formed by end-linking to quantitative conversion does not contain dangling chains, but the branch points carry remaining isocyanate functions. Consequently, the average functionality of the branch points is lower than that of the multifunctional compound B_f reacted, and the conversion of A functions (deficient) at gelation is higher than it would be for $r - 1$.

The same experimental set-up was used to measure G' and G'' as a function of reaction time, at various frequencies, and to simultaneously follow the conversion with time, by means of FTIR.

As in the previous case, the plots of log G' and of log G"
versus reaction time cross each other at a reaction time t_c - i.e.
at a conversion p_c - regardless of the frequency of the oscillatory
shear applied. Plots of log G' and log G" versus the frequency ω,
at p_c, show values of both moduli lying on the same straight line
with a slope 1/2, as in the case of stoichiometric mixtures. The
fit between calculated and experimental gel points is good.

2 - **If the A functions   are in excess**, (r < 1) dangling chains
still remain at high conversion, and the number of elastically
effective network chains is reduced. The same experimental
conditions were chosen, and the same techniques were used to follow
the gelation process.

The results obtained in this case (Figure 7) are quite
different from those discussed above. At the gel point, G" is still
higher than G', and the ratio G"/G' is found to be close to $\sqrt{3}$,
over the whole range of oscillatory shear frequencies applied. The
common exponent of the relation between G' (or G") and ω is not 1/2
(as in the preceding case), but close to 2/3. It takes 15 more
minutes for G' to become higher than G". G' tends to level off

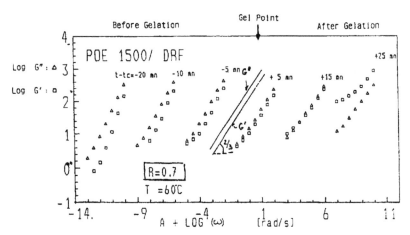

Fig. 7 - Log plots of the storage and loss moduli as a function of
the oscillatory shear applied, at various reaction times before and
after gelation. Non-stoichiometric system with r = 0.7.

still later. These experimental observations do not violate the Kramers-Kronig relations[17], which stipulate that if one dynamic modulus (G' or G") is known over the entire range of frequencies applied, the other modulus can be calculated as a function of ω.

The frequency dependence of the viscoelastic properties of non-stoichiometric systems undergoing end-linking reactions has also been investigated by Durand et al[18], near the gelation threshold. A star-shaped polymer precursor, fitted with OH functions at the outer end of the branches, was reacted with a diisocyanate to achieve the couplings. These authors confirmed that both log G' and log G" vary linearly with log ω, and that the slope of the lines is equal to 0.7.Their interpretation is based upon the analogy suggested by de Gennes[19] between the crosslinking process and a conducting random network. The value of the common exponent of the relations of G' and G" versus ω is very close to the predicted one. However, when stoichiometric conditions are used, the value of the exponent obtained experimentally is significantly different from that predicted within de Gennes' analogy, as shown by our results[14] and those of Winter[12] as well.

Recently another approach of the dynamics of crosslinking media has been proposed[20], based upon the self-similar connectivity of branched macromolecules and upon a scaling theory of fractal correlations. This treatment does not take stoichiometry as a relevant parameter. A power law is expected between the storage and loss moduli, and the frequency of the applied shear : $G' = G" \propto \omega^n$, with $2/3 < n < 1$. Again the experimental results available so far for stoichiometric systems are not in agreement with this theoretical exponent. n is found systematically lower, close to 1/2.

To interpret the results obtained, reference can be made to the predictions of Zimm[21] concerning dilute solutions of linear macromolecules, in which hydrodynamic interactions between segments -including those belonging to the same molecule - are taken into account. According to this model, at high frequencies, both G' and

G" are expected to be proportional to $\omega^{2/3}$, and to differ by a factor $\sqrt{3}$, whereas in the low frequency range the predictions of the Zimm model do not contradict those of the Rouse model : G' is proportional to ω^2 and G" to ω.

It remains to explain why stoichiometric gels behave Rouse-like, while reaction media containing precursor chains in excess would exhibit a Zimm-like behavior. In solutions of high molecular weight polymers the transition between these two states occurs upon increase of the concentration. Rouse-type behavior prevails at high concentrations whereas Zimm-type behavior is observed at lower concentrations, when hydrodynamic interactions dominate (Figure 8). At elevated concentrations, the latter are screened by inter-chain interactions. The "discrete" macromolecules which are still present at (and beyond) the gel point are assumed to play a role similar to that of the solvent in dilute polymer solutions. In order to evaluate the validity of this hypothesis, an end-linking experiment was run under stoichiometric conditions, but in the presence of zero-functional PEO chains that do not participate in the end-linking reaction. These chains exert screening, and strongly

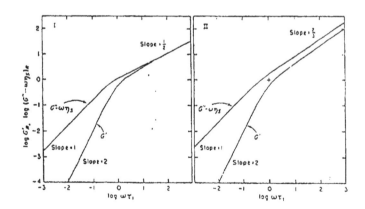

Fig. 8 - Schematic representation of the Rouse type and of the Zimm like behavior : storage and loss moduli as a function of frequency of the shear applied. (From J.D. Ferry, ref 21 page 213)

influence the rheological behaviour of the reaction medium. Despite stoichiometric conditions, a typical Zimm-like behavior was observed : when gelation has occurred G" is still higher than G' by a factor 1.7, whatever shear frequency is applied, and the common exponent of the variation of G' and G" versus ω is 2/3 instead of 1/2. The dangling chains could possibly be assigned an influence similar to that of the "discrete" macromolecules present.

CONCLUSION

There are obviously other ways to account for the observed rheological behavior of the reaction medium in which a network is formed by end-linking. But the interpretation we have put forward is self-consistent, compatible with the Kraemer-Kronig relations, and it accounts, at least qualitatively, for the observed rheological behavior of the reaction medium during network formation by end-linking.

REFERENCES

1 - G. Beinert, A. Belkebir, J. Herz, G. Hild, P. Rempp
        Faraday Disc. Chem. Soc., **57** 27 (1974)
2 - Y. Gnanou, G. Hild, P. Rempp,
        Macromolecules **17** 945 (1984) ; **20** 1662 (1987)
3 - D. Durand, F. Naveau, J.P. Busnel, M. Delsanti, M. Adam
        Macromolecumes **23** 2011 (1989)
4 - C.W. Macosko, D.R. Miller
        Macromolecules **9** 109, 206 (1976)
5 - C.W. Macosko, British Polymer J. **17** 239 (1985)
6 - J. Bastide, R. Duplessix, C. Picot, S. Candau
        Macromolecules **17** 83 (1984)
7 - J. Herz, P. Rempp, W. Borchard
        Adv. Polymer Sci. **26** 105 (1978)
8 - C.Y. Tung, P.J. Dynes
        J. Applied Polym. Physics **27** 569 (1982)

9 - M. Delsanti, M. Adam, D. Durand
                Macromolecules **18** 2285 (1985)

10 - M. Adam, M. Delsanti, J.P. Munch, D. Durand
                J. Physique **48** 1809 (1987)

11 - F. Chambon, H.H. Winter,   Polymer Bull. **13**   499 (1985)
      H.H. Winter, F. Chambon
                J. Rheology **30** 367 (1986); **31** 683 (1987)

12 - F. Chambon, Z. Petrovic, W.J. MacKnight, H.H. Winter,
                Macromolecules **19**   2146   (1986)

13 - H.H. Winter, P. Morganelli, F. Chambon
                Macromolecules **21** 532 (1988)

14 - R. Muller, E. Gérard, P. Dugand, P. Rempp, Y. Gnanou
                Macromolecules **24**   (1991)

15 - E. Franta, E. Gérard, Y. Gnanou, L. Reibel, P. Rempp
                Makromol. Chemie **191** 1689 (1990)

16 - E. Gérard, Y. Gnanou, P. Rempp,
                Macromolecules **23** 4299 (1990)

17 - see, for instance, N.W. Tschoegl "The phenomenological
                Theory of Linear Viscoelastic Behavior"
                Springer Verlag, Berlin 1989

18 - D. Durand, M.Delsanti, M. Adam, J.M. Luck
                Europhysics letters **3** (3)   297 (1987)

19 - P.G. de Gennes J. Physique **36** 1049 (1975)o

20 - J.E. Martin, D. Adolf, J.P. Wilcoxon,
                Phys. Reviews **39** 1325 (1989)

21 - see, for instance, J.D. Ferry, "Viscoelastic Properties of
                Polymers" Third Edition, Wiley, New York, 1980

Polymer Networks '91 pp. 39-62
Dosek and Kuchanov (Eds)
© VSP 1992

Comprehensive approach to the theory of polymer networks. I. Molecular theory of gelation

S.I. Kuchanov and S.V. Panyukov

Polymer Chemistry Department, Moscow State University 119899 Moscow, Russia

INTRODUCTION

For more than half a century, starting with pioneer work by Kuhn [1], the theoretical description of elasticity and some other properties of polymer networks has been drawing attention of many scientists. Several different phenomenological approaches to the quantitative interpretation of regularities observed in the behavior of polymer networks are presently known. However, their rigorous molecular theory is not available yet. The elaboration of such a theory is connected with a number of difficulties of principal character. The most important of them is, evidently, the necessity of simultaneous account of two kinds of disorder: topological and thermodynamical [2]. With respect to polymer networks this means that when calculating their thermodynamic and correlation characteristics one should take into consideration in a proper way constraints imposed on thermal motion of nodes and chains of particular network by its topological structure. The latter is formed during the network synthesis, so that some alterations of its conditions could entail an essential change of the topology of the obtained network and, consequently, result in a change of a set of its service properties.

Taking into account above arguments, when elaborating the rigorous equilibrium theory of polymer networks, one is supposed to solve, generally speaking, two problems. The first of them, referring to statistical chemistry of polymers, consists in establishing correlations between conditions of the synthesis of the network and its resulting topology. The second problem, concerning the calculation of thermodynamic and correlation characteristics of the network with given topological structure, is consistent with the statistical physics of the systems with constraints. Problems of such a kind are treated by the theory of disordered systems, methods of which (for instance, the famous "replica trick", introduced by S.Edwards [3,4] prove to be rather efficient for the description of network polymers.

In the framework of this theory they used to distinguish "quenched" and "annealed" disordered systems. For the first of them probability characteristics of topological disorder are considered to be given external parameters, while for the second kind systems they are internal parameters determined by equilibrium conditions. In macromolecular science such systems can be corresponded to those, where the equilibrium processes of polymer formation take place. In such processes the reaction mixture at any moment is considered to be in complete thermodynamic equilibrium, including that with respect to chemical reactions of the formation and decomposition of polymer molecules. If these reactions do not occur the molecular structure of polymer network remains unchanged in the course of its deformation, the interaction with low-molecular solvents as well as under the action of some other physical factors.

To develop the theory, treating experimental data on the behavior of quenched polymer network influenced by physical factors, along with final system, where the experiment is carried out, one should consider independently the initial system, where the polymer network is formed. The first who took correct account of this special feature of network polymers for the elaboration of their quantitative theory was S.Edwards. In the framework of his original treatment [3-5] the formulae of traditional statistical mechanics were extended to systems with frozen-in degrees of freedom.

The main peculiarity of our approach to the theory of polymer networks consists in comprehensive character of such a treating, which provides due account of chemical and physical factors, influencing both topological structure of a network and its mechanical properties. In this paper we shall introduce some basic ideas of such an approach and briefly review a number of fundamental results we managed to obtain in the framework of this theory.

THE STAGE OF POLYMER NETWORK FORMATION

The principal purpose of the theory treating this stage is to find quantitative correlations between the statistical characteristics of obtained branched and network polymers and the conditions of their synthesis. These are normally influenced by chemical structure and stoichiometry of initial compounds as well as by the mechanism and kinetics of chemical reactions occurring in the course of network formation. The simplest process of such a kind, which we have chosen to illustrate main ideas and results of our approach is condensation (e.g. step growth) polymerization. They usually differentiate equilibrium and nonequilibrium polycondensation, depending on the character (equilibrium or nonequilibrium) of elementary reactions between functional groups. To calculate the molecular weight distribution (MWD) of polymers we proceeded in the first and in the second case, respectively, from the Gibbs distribution and from the solution of the infinite set of kinetic equations for concentrations of molecules with given numbers of monomer units and functional groups of all kinds [6].These thermodynamic and kinetic approaches in the framework of chosen physicochemical model of polycondensation are rigorous in contrast to the widespread statistical treating, which postulates the possibility for the MWD of polymers to be derived via some speculative probability considerations. The idea of statistical approach, advanced by Flory in his early papers [7], had been later essentially refined by Gordon [8,9] who suggested to apply the mathematical apparatus of the theory of branching processes for the calculation of statistical characteristics of branched and network polymers. Since the statistical approach has a number of advantages it is important to know the areas of its applicability as well as expressions connecting formal probability parameters of branching process with thermodynamic and kinetic parameters of the reaction system.

We managed to give [10-12] the rigorous solution of the problem in question and to introduce the appropriate branching process for the description of an arbitrary polycondensation system provided the latter obeys the Flory model postulates:

1. Invariability of reactivity of all functional groups in the course of process takes place (Flory Principle)
2. There are no intramolecular cyclyzation reactions in molecules of sol (tree-like sol topology).

3. Character of spatial interactions between monomer units and molecules of solvent is not taken into account (θ-conditions).

The results [10-12] deduced in the framework of this model have been further extended [13] over two-stage processes of polymer network synthesis, when the oligomer products obtained in the first stage via cross-linking under changed conditions form the network in the second stage.

The most appropriate for the quantitative description of molecular structure of branched and network polymers is the graph theory language [14] which coupled with methods of the theory of branching processes enabled us to formalize and to solve a number of problems consistent with the calculation of topological characteristics of such polymers [15-25].

Among these characteristics of polymer network cycle rank of its molecular graph (equal by definition the smallest number of edges, which one has to delete in order to reduce this cyclic graph to tree-like one) is of primary importance. The expression for the elasticity modulus of a phantom network comprises as a factor this topological characteristic [26]. The problem of the calculation of the cycle rank of its molecular graph has been solved [27] for networks obtained for the process of ideal (e.g. obeying the Flory's postulates) polycondensation of an arbitrary mixture of monomers with any distribution for their functionality.

Along with such properties as the elasticity, which are being determined by global characteristics of polymer network topological structure, there are properties depending only on the local scale characteristics of the network topology. For instance, the glass transition temperature (T_g) of a network polymer can be calculated by means of formulae, which are sums of additive contributions of structural elements (atoms or bonds as well as their couples, triples and so on) proportional to relative fractions of such elements in the polymer [28]. Therefore, when calculating similar properties one faces the necessity to characterize quantitatively local topological structure of polymer network. The constructive approach to this problem is based on the description of such structure via consideration of the hierarchy of subgraphs U_k (k=1,2,...) as components of the molecular graph of network polymer [16,19,22]. Each of these subgraphs U_k, called "k-ada", corresponds to the network fragment consisting of k monomer units linked by chemical bonds. Such units we have suggested to differentiate not only by their chemical structure (by types), but also by their kinds [15]. The latter is determined by numbers of different chemical bonds adjacent to the given unit with regard to their configuration. Setting the distribution of monads, dyads, triads and so on we can more and more precisely characterize quantitatively the local scale of the topology (e.g. microstructure) of branched and network polymers. The fractions of these "k-ad" on the one hand can be experimentally measured via NMR-spectroscopy method, on the other hand they can be calculated in the framework of corresponding model of network formation by means of the methods of the theory of branching processes. The comparison of such theoretical results to experimental data allows one, in particular, to draw a conclusion concerning the adequacy of some model to real polycondensation system. The efficiency of this treatment has been convincingly proved [23,24] for the process of synthesis of urea-formaldehyde resins.

In the framework of developed molecular theory we managed to estimate the range of applicability of the ideal polycondensation postulates as well as to extend the theory via excluding restrictions imposed by this model.

The simplest extension of such a kind consists in taking into consideration so called "substitution effect". This effect connected with the alteration of activity of functional groups is due to steric and induction influence of chemical bonds, formed when neighbor groups in monomer unit have reacted. Natural question arises whether it is possible to describe via branching process the branched polycondensation of monomers with kinetically dependent functional groups where the Flory Principle is not implemented. Prior to the publication of our papers [10,29] there was an opinion in the literature that the answer to this question is always positive. Our rigorous (in the framework of the model of "substitution effect") thermodynamic [10,30,15] and kinetic [10,29,18] consideration has shown that above conclusion turns out to be correct only for the products of equilibrium polycondensation. However, under non-equilibrium regime of its performing, there are systems where molelcular structure distribution (MSD) of formed tree-like molecules, generally speaking, can not be described by any branching process. Nevertheless for some systems of such a kind we have suggested an algorithm [18] for the construction of hierarchical sequence of branching processes permitting one with increasing accuracy to provide an approximate statistical description of polymer products.

When elaborating theories, which take into consideration intramolecular cyclization and/or physical interactions of monomer units between each other and with the solvent, it is obviously indispensable along with chemical structure of macromolecule to take also account of their conformations, e.g. mutual location of monomer units in the space. In terms of this treating polymer system is not considered any longer as statistical set of abstract graphs, but as a set of graphs embedded in three-dimensional space. The simplest way to realize such a consideration implies the usage of lattice models of the gel-formation and, particularly, the percolation model [31,32] as the most widespread among them. In the framework of this models all sites of the lattice are supposed to be occupied by nodes of molecular graphs, while the bonds on lattice can be either occupied or not by edges of these graphs depending on the chemical reaction occurring between functional groups of monomer units corresponding to particular nodes. Although above model due to its simplicity seems to be rather attractive it, however, does not provide sufficiently adequate account of many concrete peculiarities of polymer system under consideration. The molecular theory, we are developing, does not proceed from the lattice model but from the continuum one. This permits us to express all the unknown sole and gel characteristics through comparatively small number of having obvious meaning chemical and physical parameters, which can be easily obtained from data of simple experiments.

For the elaboration of comprehensive theory of polymer networks it is natural to consider firstly those of them, which are formed under the equilibrium conditions. The second part of this paper is devoted to brief review of the fundamental results we have obtained in the framework of such a theory with respect to branched and network polymers obtained via the process of equilibrium polycondensation. To describe equilibrium systems along with the apparatus of branching processes theory we applied the methods of the field theory which are the most promising tool among modern approaches.

CHEMICALLY-EQUILIBRIUM FORMATION OF NETWORK POLYMERS

Developing quantitative theory of such a process one faces three fundamental problems. The first of them consists in finding molecular structure characteristics of equilibrium polycondensation products, while the second and the third, respectively, are connected with the calculation of thermodynamic and correlation characteristics of the system.

For the solution of the first among mentioned problems serious difficulties arise concerning the correct account of intramolelcular cyclization reactions occurring in sole. For those systems, where the contribution of the latter is rather small in comparison with that of intramolecular reactions of functional groups we have advanced an approach [17,20], enabling one to give up the second of three Flory's postulates of the ideal polycondensation model. In the framework of the simplest version [17] of this approach only trivial cycles (whose cycle rank equals, by definition, unity) are assumed to be in sol molecules. Each molecular graph in this case is "cactus", which can be characterized by the set of numbers of cycles with different size and nodes that enter no cycles. It was shown [17] that to these fragments of molecular graphs one can correspond different type particles of a certain branching process, which allows one to obtain rigorous (in the framework of the model under consideration) expression for molecular structure characteristics of sole and gel.

In further paper [20] to find these characteristics authors derived formally precise expressions with the account of the possibility of formation of cycles with arbitrary topology, which correspond to the particles of

different types of the generalized branching process. It
is essentially that these expressions look like
expansions in powers of parameter ε, which is small in
systems where intramolecular cyclization is weak
enough.In the zeroth and first order in parameter ε one
obtains, respectively, formulae of the ideal
polycondensation theory and the results of previous
paper [17]. Each of subsequent items of such an
expansion being the coefficient of ε^r takes into
consideration the contributions of cycles with cycle
rank r. The increase of the number of items of this
infinite series, which is retained when we truncate it,
allows us to take account of the contributions of the
cycles with more and more complicated topology,
improving in such a way the the order of accuracy of
obtained approximate results.

THE FIELD THEORETIC FORMALISM

The field theory methods seem to be the most
efficient among those applied in the theory of gelation
and polymer networks formation. They provide a
successive account both of volume physical interactions
between units and chemical interactions between their
functional groups. It also worth emphasizing that the
field theoretic formalism gives one a possibility in the
framework of a single approach to find the solution to
two different problems. The first of them is connected
with molecular structure description of the formed
polymer while the second one is consistent with the
calculation of thermodynamic potentials of reaction
systems [33,27,22]. The solution to the last problem is,
obviously, of great practical value since it enables one
to formulate conditions of phase separations occurring
during polymer network formation.

Besides in in terms of the approach proposed it is possible to find correlation functions $\theta_{\alpha\beta}(r'-r'')$ of fluctuations of microscopic densities $\rho_{\alpha}^{m}(r')$ and $\rho_{\beta}^{m}(r'')$ of monomer units R_{α} and R_{β} located at points r' and r''. In the simplest case of homopolycondensation of monomer RA^{f} with f identical functional groups A, considered further as example, the matrix $\theta_{\alpha\beta}(r)$ is reduced to scalar function $\theta(r)$. Its Fourier transform $\tilde{\theta}(q)$ can be obtained experimentally from the data on angular dependence of the amplitude of the light or neutron scattering, the condition $\tilde{\theta}(0)=\infty$ corresponding to the moment when the system reaches the spinodal.

Along with overall density of monomer units the density of units of separate polymer molecule is also considered. Averaged over all such molecules of the system under consideration the correlator $\chi(r)$ of their density fluctuations can be denominated the "pair connectedness" function by analogy with the percolation theory [34], where it has an identical meaning of the probability to find a pair of sites of some cluster, located at a distance r one from another. The radius of polymer molecules as well as their other geometric characteristics can be easily calculated, provided the function $\chi(r)$ is known.

Two Generating Functionals (GF)s contain an exhaustive information about the system under consideration. The first of them, $-\Omega\{h\}/T$, being GF of correlators of overall units density is up to a factor (-1) none other than divided by the temperature T the thermodynamic Ω-potential for the system under the action of an external field H(r). This potential enables one to find conditions of phase transition, while its second variation derivative with respect to external field equals overall density fluctuations correlator

$$\theta(r'-r'') = -T \left. \frac{\delta^2 \Omega\{H\}}{\delta H(r')\delta H(r'')} \right|_{H(r)\equiv 0} \tag{1}$$

Analogous second derivative, but now with respect to dummy variable $s(r)$ of GF $\Psi\{s\}$ of correlators of monomer units density of individual polymer molecules just gives the "pair connectedness" function

$$\chi(r'-r'') = \left. \frac{\delta^2 \Psi\{s\}}{\delta \ln s(r')\delta \ln s(r'')} \right|_{s(r)\equiv 1} \tag{2}$$

The divergence of its Fourier transform $\tilde{\chi}(q)$ at the point $q=0$ corresponds to the appearance in the reaction system of infinite polymer network of gel. It is worth underlining that the first variation derivative of GF $\Psi\{s\}$ with respect to $\ln s(r)$ at $s(r)=s$ equals generating function (gf) $G_W(s)$ of weight MWD $f_W(1)$ of polycondensation products

$$G_W(s) \equiv \sum_{1=1}^{\infty} f_W(1)s^1 = \left. \frac{1}{\rho} \frac{\delta\Psi\{s\}}{\delta \ln s(r)} \right|_{s(r)=s} \tag{3}$$

Therefore the task of elaboration of the gelation theory is reduced to the construction of two mentioned functionals $\Omega\{H\}$ and $\Psi\{s\}$ in the framework of corresponding molecular model as adequate to real system as possible.

The Lifshitz–Erukhimovich model [22] obviously meets these requirements. Within the framework of this model the probability $P(G_N\{r_1\})$ of any state of reaction system (characterized by its graph $G_N\{r_1\}$, whose N vertices are located at points r_1, $r_2,\ldots r_1,\ldots r_N$ of three-dimensional space) is determined by relation

$$P(G_N\{r_i\}) = P^{(1)}P^{(2)}; \quad P^{(2)} = \prod_{(ij)} L\,\lambda(r_i - r_j)$$

$$P^{(1)} = M^N \exp\left\{\left[\Omega - \sum_{i=1}^{N} H(r_i) - \sum_{(ij)} V(r_i - r_j)\right]/T\right\}$$

(4)

This formula is just the Gibbs distribution of the system with monomer units as components, which interact between each other with the potential V and are located in the external field H. Their activity M along with equilibrium constant L of elementary reaction between functional groups are ordinary thermodynamic parameters.

The expression for probability (4) is the product of two factors. The first of them $P^{(1)}$, taking into account only physical spatial interactions of units, looks like the Gibbs distribution of regular system, consisting of small molecules only. Polymer specificity of the system under consideration is due to the second factor $P^{(2)}$, which is equal to the product of factors $L\lambda(r_i - r_j)$ over all edges of a graph $G_N\{r_i\}$. Each of these factors is proportional to the probability $\lambda(r_i - r_j)$ to find a pair of chemically linked monomer units at a distance $r_i - r_j$ one from another. The function λ accounts for the conformation entropy decrease as a result of the constraints for mobility of a pair of units, caused by the chemical bond between them.

In terms of this model it is possible to derive exact expression for the thermodynamic Ω-potential of the system located in external field H(r)

$$-\frac{\Omega\{H\}}{T} = \ln\left\langle \exp\left[-\frac{\Omega_{su}\{\hat{z}\}}{T}\right]\right\rangle_\varphi$$

(5)

$$\hat{z}(r) = z(r)[1+\varphi(r)]^f, \text{ where } z(r) = M \exp[-H(r)/T] \quad (6)$$

As it follows from expression (5) the grand partition function of the polymer system is the result of averaging designated   by angle brackets) over some random Gaussian field $\varphi(r)$ with the Lagrangian $L\{\varphi\}$

$$L\{\varphi\} = \frac{1}{2L} \iint \Lambda(r'-r'')\varphi(r')\varphi(r'')dr'\,dr''$$

$$(7)$$

$$\int \Lambda(r'-r)\lambda(r-r'')dr = \delta(r'-r'')$$

of the partition function of more simple system of monomer units without chemical bonds but being under the action of additional random field $-fT\ln(1+\varphi)$. For such a kind traditional "system of separate units" one can use expressions for thermodynamic potential $\Omega_{su}\{z(r)\}$, obtained before in the framework of well-known models ("lattice gas" model, for instance), substituting independent variable $z(r)$ for $\hat{z}(r)$ in this expression (6). It permits one to take correct account of excluded volume and the van der Waals attraction as well as to calculate both thermodynamic and correlation characteristics of polymer system.

Final relationships for the Gibbs free energy and correlator of overall density fluctuations (1) can be directly used to compare theoretical conclusions with experimental data. Having calculated phase diagram it is possible particularly to point out values of chemical and physical parameters, where phase separation occurs before gelation and as a result unhomogeneous network is formed according to microgelation mechanism.

To find GF $\Psi\{s\}$ of correlators of monomer units density fluctuations of individual polymer molecules one has to solve essentially more complicated problem. The latter is really rather combinatorial problem than thermodynamic one and it is similar to popular problem of statistic description of percolation clusters. However we managed to reduce it to the solution of purely thermodynamic problem of calculating Ω-potential for some abstract system of "colored units". In such a system every monomer unit is supposed to have n additional copies, each of them being specified by its own "color", denoted by index i=0,1,...,n. In this model each particle of color i, except "white" i=0, is acted upon parallel with real external field H(r) by additional virtual one $H_i^{vir}(r) = -T \ln s(r)$ and besides only groups of an identical color can react between each other forming chemical bond. So calculating Ω-potential of the system of colored units (Ω_n) and "erasing" subsequently their colors it is possible to get the unknown GF $\Psi\{s\}$. Such "color erasing" procedure implies that one has to put value of variable n equal zero in Ω_n, however having previously differentiated it with respect to n

$$\Psi\{s\} = - \left. \frac{d\Omega_n}{Tdn} \right|_{n=0} \tag{8}$$

The thermodynamic potential Ω_n can be immediately written down in accordance with general rules [27] which we have formulated for thermodynamic description of the products of equilibrium polycondensation of arbitrary mixture of monomers. In the case under consideration of monomer RA^f homopolycondensation for potential Ω_n the following formula

$$- \frac{\Omega_n\{H\}}{T} = \ln \left\langle \exp \left[- \frac{\Omega_{su}\{\hat{z}^{(n)}\}}{T} \right] \right\rangle_\phi \tag{9}$$

is true and analogous to formula (5) but differing from it by two peculiarities. The first of them is that the averaging in the expression (9) is carried out over (n+1)-component random replica field $\vec{\phi}$, probability measure on realizations of which is given by the Lagrangian

$$L_n\{\vec{\phi}\} = \sum_{i=0}^{n} L\{\varphi_i\}; \quad \vec{\phi} = (\varphi_0, \varphi_1, \ldots, \varphi_i, \ldots, \varphi_n) \quad (10)$$

where the functional $L\{\varphi_i\}$ is determined via formula (7). The second peculiarity consists in the fact that the expression (9) has the function $\hat{z}^{(n)}(r)$ instead of $\hat{z}(r)$ (6) as an argument of the functional Ω_{su}

$$\hat{z}^{(n)}(r) = M \sum_{i=0}^{n} \exp\left[-H_i(r)/T\right] \left[1 + \varphi_i(r)\right]^f \quad (11)$$

$$H_i(r) = H(r) + H_i^{vir}(r), \quad H_i^{vir}(r) \equiv (\delta_{i0} - 1)T \ln s(r) \quad (12)$$

where δ_{i0} is the Kroneker delta-symbol.

We can arrive at an interesting conclusion having previously put GF (8) into formula (2) and having permuted the order of differentiation of the functional Ω_n with respect to virtual field and n. As it follows from the comparison of the obtained expression to the formula (1), in order to find the pair connectedness function it is sufficient to get the corresponding correlator of overall density fluctuations in the system of colored units and then to "erase" their colors differentiating this correlator with respect to n at n→0. It is noteworthy that the gel point (in which, by definition, the condition $\tilde{\chi}(0) = \infty$ is implemented by definition with the growth of conversion) can be corresponded to the second order phase transition in the system of colored units where the divergence of the zeroth Fourier transform of their density correlator appears at this point.

SELF-CONSISTENT FIELD (SCF) APPROXIMATION

This approximation enables one to simplify essentially foregoing expressions (5-12), which are formally precise in the framework of the model under consideration. The SCF approximation corresponds to the evaluation of functional integrals (5) and (9) via the steepest descent method. Above procedure with respect to the first of these integrals permits one to find expressions for thermodynamic potentials of the space homogeneous system and to write down its equation of state

$$P(\rho) = P_{su}(\rho) - \frac{1}{2} Tf p \rho, \quad \frac{p}{f\rho(1-p)^2} = L \quad (13)$$

which expresses the dependence of pressure P on the density ρ through similar dependence $P_{su}(\rho)$ but referring to the system of "separate units", which for "lattice gas" model is as follows

$$\frac{P_{su}(\rho)v}{T} = -\ln(1-v\rho) - \frac{\gamma}{2T}(v\rho)^2 \quad (14)$$

Here the particular gel formation model comprises three independent parameters: close packed volume v of monomer unit, parameter γ characterizing the energy of pair interaction of such units and the constant of equilibrium L the reaction between functional groups. The second of equations (13), which is none other than mass action low of the mentioned reactions, allows one to find the value of conversion p of these groups. It worth noting that their chemical binding is equivalent (in accordance with equation (13)) to the effective attraction of units and can induce a phase transition even under conditions when the latter is absent in the system of separate units. Proceeding from the equations (13) one can derive the expression

$$p_{sp} = \frac{1}{\nu f - 1} \; ; \quad \frac{1}{\nu} \equiv \frac{1}{T} \frac{\partial P_{su}}{\partial \rho} = \frac{1}{1 - \nu\rho} - \frac{\gamma}{T} \nu\rho \qquad (15)$$

for conversion p_{sp} at the spinodal. In the absence of spatial interactions between units, when

$$P_{su}(\rho) = T\rho, \quad \nu \equiv 1, \quad p_{sp} = 1/(f-1) \qquad (16)$$

it turns out that in the framework of the Mean Field Approximation the spinodal coincides with the gel point $p^* = 1/(f-1)$. With an account of such interaction, depending on their character, gelation as it ensues from relations (15) can either precede or follow the phase separation. In the first case macrogelation occurs, while in the second one microgelation takes place.

Whereas to derive the equation of state for spatially homogeneous system one can ignore completely the deviations of all fields from its mean values, for the calculation of correlators $\theta_{\alpha\beta}(r'-r'')$ there is a necessity to take into account spatial fluctuations of monomer units density. In a simplest way such an account used to be performed in the framework of the Random Phase Approximation (RPA) or some analogous approaches [35] resulting in the same final expressions for $\theta_{\alpha\beta}(r'-r'')$. These expressions are also easy to derive in terms of SCF approximation which provides due account of the dependence of extreme configuration of the self-consistent field on spatial co-ordinates. In particular, a simple formula can be obtained for the Fourier transform of the correlator (1)

$$\tilde{\theta}^{-1}(q) = \tilde{g}^{-1}(q) - C; \text{ where } C = v\left[\frac{\gamma}{T} - \frac{1}{1-\nu\rho} \right] \qquad (17)$$

Hereafter $\tilde{g}(q)$ - the Fourier transform of "structure function" $g(r'-r'')$

$$\tilde{g}(q) = \rho\left[1 + \frac{p f}{\tilde{\lambda}^{-1}(q) - p(f-1)}\right]; \quad \tilde{\lambda}(q) = e^{-a^2 q^2} \quad (18)$$

which is the density correlator in a system with no spatial interactions, whose contribution is taken into account in formula (17) through "direct correlation function" C of the "system of separate units". A Gaussian distribution with dispersion a^2, which can be chosen as the function $\lambda(r)$, has the Fourier transform $\tilde{\lambda}(q)$, adduced in formula (18).

In order to describe gelation as well as the MWD of polymers one should consider the GF $\Psi\{s\}$, i.e. to find the thermodynamic potential $\Omega_n\{\tilde{H}\}$ (9) in the system of colored units and then to "erase" their colors in accordance with expression (8). When calculating functional Ω_n by means of the steepest descent method we will obtain the set of equations for the calculation of the SCF, which at low conversion has the sole solution, symmetric with respect to permutations of colors. As the conversion p increases attaining a critical value p^* the colored units system undergoes the second order phase transition when additional n asymmetric solutions appear as a result of the bifurcation. They describe the gel in contrast to symmetric with respect to replicas (i.e. colors) solution describing sol-fraction. Correlators of sol $\Psi^{(s)}\{s\}$ and gel $\Psi^{(g)}\{s\}$ depend only on symmetric and asymmetric solutions, respectively.

To calculate both structure and conformation characteristics of polymer molecules forming the sol-fraction it is convenient to use the functional

$$G_W(r,\{s\}) = \frac{s(r)}{\rho}\frac{\delta\Psi^{(s)}\{s\}}{\delta s(r)} = s(r)\,\Xi^f(r,\{s\})$$

$$\Xi(r,\{s\}) \equiv 1-p+p\int\lambda(r-r')U(r',\{s\})dr'\,,\qquad(19)$$

$$U(r',\{s\}) \equiv s(r')\Xi^{f-1}(r',\{s\})$$

This relations are easy to reduce via setting $s(r)=s=$const into well-known expressions [8,9]

$$G_W(s) = s\xi^f(s),\ \ \xi^{(s)} \equiv 1-p+pu(s),\ \ u(s) \equiv s\xi^{f-1}(s)\quad(20)$$

for gf (3) of MWD of ideal polycondensation products.

Formulae (19) provide rather interesting interpretation and can be easily derived [36] by means of the theory of branching processes. However we suggested to apply for the calculation of the statistical characteristics of polymers not ordinary but general branching processes. They describe parallel with "reproduction" and "death" of particles (as in case of traditional statistic treatment [8,9]) also the diffusion of this particles in space.

Differentiating the expression (19) with respect to the function $\ln s(r')$ and setting $s(r)=1$ we arrive at equation for the pair connectedness function (2). The solution to this equation can be easily obtained when the Fourier transform is applied

$$\tilde{\chi}^{(s)}(q) = \rho\xi^f + \frac{\rho pf\xi^{2f-2}}{\tilde{\lambda}^{-1}(q)-p(f-1)\xi^{f-2}} = \tilde{g}^{(s)}(q)\qquad(21)$$

where the dependence $\xi\equiv\xi(1)<1$ on conversion p can be derived from equations (20). In a similar way, having calculated the second variation derivative (2) of $\Psi^{(g)}\{s\}$ we can get the expression for the Fourier transform of pair connectedness function of gel, which coincides with the correlator of density fluctuations of its units

$$\tilde{\chi}^{(g)}(q) = \tilde{g}^{(g)}(q) + \tilde{g}^{(g)}(q)\tilde{\beta}(q)\tilde{g}^{(g)}(q) = \tilde{\theta}^{(g)}(q) \qquad (22)$$

where the following designations are used

$$\tilde{\beta}^{-1}(q) \equiv C^{-1} - \tilde{g}(q), \quad \tilde{g}^{(g)}(q) \equiv \tilde{g}(q) - \tilde{g}^{(s)}(q) \qquad (23)$$

According to contemporary molecular theory of phase transitions SCF approximation will be true when average fluctuations of gel density on scale of correlation length are rather small. In very close vicinity of gel point this condition is certainly broken. However, in systems, where value of the Ginzburg parameter

$$Gi = \left[\frac{(f-2)^2}{16\pi f(f-1)\rho a^3}\right]^{2/3} \sim \varepsilon^{2/3} \qquad (24)$$

is small enough, such a vicinity is too narrow to be accessible to experimental investigations. In this case one can suppose that SCF approximation holds over whole range of conversions and obtained networks have classical topological structure. The latter sometimes can be quite different from classical one and similar to the structure of infinite cluster, which is formed in a percolation system at the vicinity of percolation threshold. Such critically branched networks of percolation type are obtained in systems, where the parameter Gi (24) value as well as ε is not small and intramolelcular cyclization reactions play important role already at the stage of the synthesis preceding polymer network formation.

In present paper we have briefly put forth only main ideas of our quantitative theory of gelation in equilibrium systems. More detailed explanation of this theory and its general results, treating a polycondensation of arbitrary mixture of monomers can be found in our review [22].

REFERENCES

[1] W.Kuhn, Kolloid Z. v.76, 258 (1936)

[2] J.M.Ziman, Models of Disorder. Cambridge Univ.Press, Cambridge-London-New York (1979)

[3] S.F.Edwards, Statistical Mechanics of Polymerized Materials, in: Proc. 4th Internat.Conference on Amorphous Materials. R.W.Douglas, B.Ellis (Eds.), Ch.30, pp.279-300. J.Wiley Publ., New York (1970)

[4] S.F.Edwards, Statistical Mechanics of Rubber, in: Polymer Networks: Structural and Mechanical Properties. A.J.Chompff, S.Newman (Eds.), pp.85-103, Plenum Press, New York (1971)

[5] R.T.Deam, S.F.Edwards, Phil.Trans.Royal Soc.(London) Ser.A, v.280, 317 (1976)

[6] S.I.Kuchanov, Methods of Kinetic Calculations in Polymer Chemistry (in Russian), Khimia Publ., Moscow (1978)

[7] P.J.Flory, Principles of Polymer Chemistry, Cornell Univ. Press, Ithaca - New-York (1953)

[8] M.Gordon, Proc.Royal Soc. (London), Ser.A, v.268, 240 (1962)

[9] G.R.Dobson, M.Gordon, J.Ch.Phys., v.41, 2389 (1964)

[10]S.I.Kuchanov,Dokl.Acad.Nauk(USSR),v.229, 136 (1976)

[11]S.V.Korolev,S.I.Kuchanov,M.G.Slinko, Dokl.Acad.Nauk (USSR), v.258, 1157 (1981)

[12]S.V.Korolev,S.I.Kuchanov,M.G.Slinko, Dokl.Acad.Nauk (USSR), v.263, 633 (1982)

[13]S.I.Kuchanov, Vysokomolek.Soed., Ser.B, v.29, 671 (1987)

[14]M.Gordon, W.B.Temple, The Graph-like State of Matter and Polymer Science, in: Chemical Applications of Graph Theory, A.T.Balaban (Ed.), pp.299-332, Academic Press, New York-London (1976)

[15]S.V.Korolev,S.I.Kuchanov,M.G.Slinko, Dokl.Acad.Nauk (USSR), v.262, 1422 (1982)

[16]S.I.Kuchanov, S.V.Korolev, M.G.Slinko, Vysokomolek. Soed., Ser.A, v.26, 263 (1984)

[17]S.I.Kuchanov, S.V.Korolev, M.G.Slinko, Vysokomolek. Soed., Ser.A, v.24, 2160, 2170 (1982)

[18]S.I.Kuchanov, E.S.Povolotskaya, Vysokomolek.Soed., Ser.A, v.24, 2179, 2190 (1982)

[19]S.I.Kuchanov, S.V.Korolev, M.G.Slinko, Polymer J., v.15, 775, 785 (1983)

[20]S.I.Kuchanov,S.V.Korolev,M.G.Slinko, Dokl.Acad.Nauk (USSR), v.267, 122 (1982)

[21]S.I.Kuchanov, S.V.Korolev, The New Approaches to the Statistical Theory of Polymer Networks Formation, in: Proc.International Conference "Rubber-84", Moscow (1984)

[22]S.I.Kuchanov, S.V.Korolev, S.V.Panyukov, Graphs in Chemical Physics of Polymers, in: Adv.Chem.Phys., v.72, 115 (1988)

[23]S.V.Korolev, S.G.Alekseeva, S.I.Kuchanov et al. Vysokomolek.Soed., Ser.A, v.29, 2378 (1987)

[24]S.V.Korolev, S.I.Kuchanov, Vysokomolek.Soed., Ser.A, v.29, 2387 (1987)

[25]S.I.Kuchanov, S.V.Korolev, Vysokomolek.Soed., Ser.A, v.29, 2309 (1987)

[26]P.J.Flory, Proc.Roy.Soc.(London), Ser.A, v.351, 351 (1976)

[27]S.I.Kuchanov, Dokl.Acad.Nauk (USSR),v.294, 342(1987) (Sov.Phys.Doklady, v.32, 392 (1987))

[28]V.Bellenger, J.Verdu, E.Morel, J.Pol.Sci., Ser.B,
    v.25, 1219 (1987)

[29]S.I.Kuchanov, Dokl.Acad.Nauk (USSR),v.249,899 (1979)

[30]E.B.Brun, S.I.Kuchanov, Vysokomolek.Soed., Ser.A,
    v.21, 1393 (1979)

[31]P.G.DeGennes, J.de Phys., v.36, 1049 (1975)

[32]D.Stauffer,J.Chem.Soc., Faraday II,v.72, 1354 (1976)

[33]S.V.Panyukov, Zh.Eksp.Teor.Fiz., v.88, 1795 (1985)
    (Sov.Phys.JETP, v.61, 1065 (1985))

[34]J.W.Essam, Reports Progr.Phys., v.43, 833 (1980)

[35]T.A.Vilgis, M.Benmouna, H.Benoit, Macromolecules,
    v.24, 4481 (1991)

[36]S.I.Kuchanov, Dokl.Acad.Nauk(USSR),v.294,633 (1987)
    (Doklady Phys.Chem., DKPCAG, v.294, 514 (1987))

Polymer Networks '91 pp. 63-78
Dosek and Kuchanov (Eds)
© VSP 1992

The role of entanglements for the mechanical behaviour of polymer networks

O. Kramer

Department of Chemistry, University of Copenhagen, Universitetsparken 5, DK-2100 Copenhagen, Denmark

ABSTRACT

The topological structure of entangled chains is incompletely trapped in lightly cross-linked elastomers, giving rise to large mechanical losses when the chains disentangle. For the initial stages of cross-linking, the longest relaxation times are much longer than for the uncross-linked linear polymer. Work by Ferry and coworkers on model networks has shown that unattached linear chains present in a network disentangle at approximately the same rate as in the uncross-linked linear polymer. The exceedingly long relaxation times are explained by the slow disentanglement of very long dangling chain ends. Well cross-linked elastomers exhibit practically no mechanical losses since the structure of entangled chains is nearly completely trapped. The contribution of entanglements to the modulus at elastic equilibrium is discussed in terms of 1,2-polybutadiene networks cross-linked in the strained state. The entanglement contribution is found to be much larger than the contribution from chemical cross-links and approximately equal to the rubber plateau modulus of the uncross-linked polymer.

INTRODUCTION

It is well established that uncross-linked flexible polymer chains are highly coiled and form a temporary rubber-like network above the glass transition temperature, T_g. At times shorter than the time to disentangle, τ_{dis}, the uncross-linked polymer has properties which are typical of cross-linked elastomers with an elongation at break of several hundred percent and a shear modulus which typically is in the range from 2×10^5 to 2×10^6 Pa [1]. The disentanglement time is found to be proportional to molecular weight to a power of about 3.4 [2]. It means that the rubber plateau region may extend over many decades of time or frequency for polymers of very high molecular weight [3]. An example can be seen in Figure 1 where the lower curve shows the rubber plateau region of uncross-linked 1,2-polybutadiene with a molecular weight of 291,000 [4]. The plateau would have been more pronounced in a double logarithmic plot of storage modulus versus frequency [3].

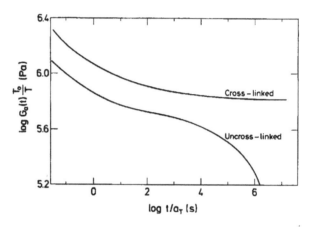

Figure 1. Stress relaxation curves for cross-linked and uncross-linked polybutadiene with 88% 1,2-structure and a weight average molecular weight of 291,000. The reference temperature is 263 K, i.e., T_g + 8 K. The cross-linked sample has about 60 cross-linked points per chain (Results of Batsberg and Kramer [4] reproduced with permission from Elsevier Applied Sci.)

By direct comparison to an ideal rubber, an apparent molecular weight between entanglement couplings is calculated from the rubber plateau modulus, G_N^o, using the affine theory of rubber elasticity [5,6]

$$G_N^o = (\frac{\rho}{M_e})RT \qquad (1)$$

where ρ is the density, M_e is the apparent molecular weight between entanglements, R is the gas constant and T is absolute temperature. The structural significance of M_e is still obscure, and it must be stressed that M_e presently is nothing more than the strand molecular weight of an ideal rubber with the same density and a modulus equal to the rubber plateau modulus of the uncross-linked polymer.

The uncross-linked linear polymer molecules disentangle completely at times longer than τ_{dis}. This makes the modulus drop towards zero and gives rise to a large peak in the loss modulus. The mechanism of disentanglement is thought to be the reptation mechanism proposed by deGennes [7-10]. However, disentanglement by the simple reptation mechanism is easily prevented by cross-linking. The polymer molecules gradually become linked together and eventually form one giant molecule at higher degrees of cross-linking. The upper curve of Figure 1 shows well cross-linked 1,2-polybutadiene with approximately 50-100 cross-linked points per polymer chain [4,11]. It can be seen that the terminal zone has disappeared and that elastic equilibrium is reached rather quickly. The absence of mechanical losses in well cross-linked elastomers indicates directly that the elastic contribution from chain entangling has been made permanent by the introduction of a large number of chemical cross-links. The observed mechanical losses in lightly cross-linked elastomers are a measure of that part of the elastic contribution which is not of a permanent nature.

Relaxation times much longer than τ_{dis} of the uncross-linked polymer appear at the initial stages of cross-linking but the longest relaxation times decrease rapidly with increasing degree of cross-linking [12]. Stress relaxation measurements made by Chasset and Thirion on natural rubber with varying degrees of cross-linking are shown in Figure 2. Elastic equilibrium is reached in a few hours for the sample with the highest degree of cross-linking. The sample with the lowest degree of cross-linking is still decades away from elastic equilibrium at the end of the experiment, i.e., after more than one day.

The first part of the paper gives a discussion of the effects of different types of network defects on the non-equilibrium properties, focusing on work by Ferry and coworkers. The second part gives a discussion of the contribution of entanglements to the equilibrium modulus of well cross-linked networks, focusing on work from our own laboratory.

NON-EQUILIBRIUM PROPERTIES

Both linear polymer molecules and branched structures exist at the initial stages of cross-linking. Studies on special networks have shown that dangling chain

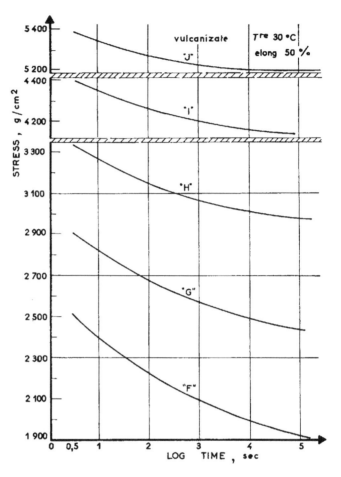

Figure 2. Stress relaxation of natural rubber networks with varying degrees of cross-linkin;
The values of the equilibrium modulus, log (G_e/Pa), are as follows: (F) 5.29; (G) 5.41; (H
5.51; (I) 5.65; (J) 5.75 (Results of Chasset and Thirion [13] reproduced with permission from
North Holland)

ends can account for the exceedingly long relaxation times observed in lightly
cross-linked polymers. Ferry and coworkers first studied different types ‹
covalently cross-linked networks containing unattached linear polymer chains
Sulphur cross-linked butylrubber with linear polyisobutylene chains [14]
sulphur cross-linked EPDM rubber with linear ethylene-propylene copolyme
chains [15]; end-linked polydimethylsiloxane networks with linear polyd
methylsiloxane chains [16]; and end-linked polybutadiene networks with linea
styrene-butadiene copolymer chains [17].

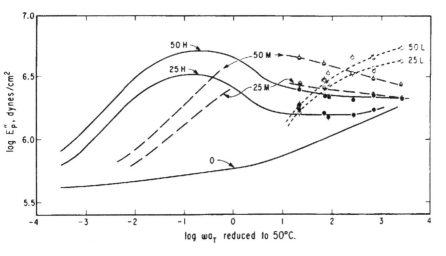

Figure 3. Loss modulus of EPDM networks containing unattached linear ethylene-propylene copolymer chains of varying molecular weights and concentrations of 0, 25 and 50%. Weight average molecular weights of the unattached chains are: (L) 36,000; (M) 167,000; (H) 450,000 (Results of Kramer, Greco, Ferry and McDonel [15] reproduced with permission from John Wiley)

Figure 3 shows the loss modulus of EPDM networks containing unattached ethylene-propylene copolymer chains of three different molecular weights and concentrations of 0, 25, or 50 %. Disentanglement of unattached chains of a particular molecular weight gives rise to a loss peak, whose location on the frequency scale is proportional to molecular weight to a power of 3-3.5, i.e., in good agreement with the prediction of deGennes [7,15]. Unattached polymer chains dissolved in a network simply disentangle at the same rate as they do in the pure melt [14] unless the network is very tightly cross-linked [16]. These experiments therefore clearly demonstrate that unattached linear polymer chains cannot be responsible for the exceedingly long relaxation times observed at the initial stages of cross-linking of long linear chains.

It has proven difficult to make covalently cross-linked networks containing unattached star-shaped polymer molecules. Kan, Fetters and Ferry [18] instead used a styrene-butadiene-styrene triblock copolymer as the host network. Figure 4 shows that 3- and 4-arm stars relax much more slowly than a linear chain of approximately the same molecular weight. One could argue that the comparison ought to be made on the basis that the arm molecular weight should be equal to one half of the molecular weight of the linear polymer chain. This would have made the difference even more pronounced.

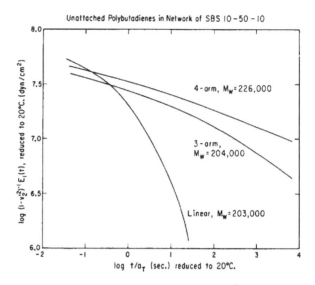

Figure 4. Stress relaxation of unattached linear and star polybutadiene molecules in a network of styrene-butadiene-styrene block copolymer at a reference temperature of 20 °C. The relaxation of star shaped molecules, which is much slower than that of linear chains, corresponds to the relaxation of dangling chain ends (Results of Kan, Ferry and Fetters [18] reproduced with permission from American Chemical Society)

Cohen and Tschoegl [19] and Kamykowsky, Ferry and Fetters [20] used styrene-butadiene-styrene triblock copolymers to study the relaxation of dangling chains by adding styrene-butadiene diblock copolymer to the triblock copolymer. Short dangling chains relax very fast which can explain the rapid equilibration of end-linked networks, even in the presence of a substantial fraction of dangling chain ends [4]. Long dangling chain ends relax so slowly that they can account for the exceedingly long relaxation times observed in very lightly cross-linked elastomers.

Substantial progress has also been made in the development of theory. It was pointed out by deGennes already in 1975 that dangling chains cannot disentangle by the simple reptation mechanism. He proposed instead that dangling chains should relax by a much slower back-tracing mechanism which would make the relaxation time depend exponentially on molecular weight of the dangling chain [21]. Pearson and Helfand developed a theory for the viscoelastic properties of a melt of star-shaped polymer molecules [22], and Curro, Pearson and Helfand expanded the treatment to the relaxation of dangling chain ends in a randomly cross-linked network [23].

THE ROLE OF ENTANGLEMENTS AT ELASTIC EQUILIBRIUM

Whether entanglements also contribute to the mechanical properties at elastic equilibrium is a question which has been difficult to answer unequivocally [24-28], the main reason being different answers from different types of experiments. The answer is of great consequence, considering the use of elastic modulus and swelling for determining the degree of cross-linking and extent of reaction in the formation of polymer networks. It would also be expedient to know whether future attempts to develop improved theories of rubber elasticity should be based on the presence of an entanglement term, or not.

Formally we may write

$$G_{total} = G_X + G_N \qquad (2)$$

where G_{total} is the measured shear modulus; G_X is the contribution from chemical cross-links; and G_N is the contribution from entanglements. The constrained junction theory developed by Ronca and Allegra [29], Flory [30], and Flory and Erman [31,32] is based on the assumption that entanglements play no role other than to reduce the fluctuations of the chemical cross-links, i.e., $G_N=0$. Flory was convinced that this assumption is correct since a large number of experimental studies on end-linked networks indicated that no entanglement term was required [24,27]. It is unfortunate that most of the work was made on polydimethylsiloxane networks for which G_N should be expected to be low in any case because of the low rubber plateau modulus of polydimethylsiloxane [1]. Even so, Gottlieb, Macosko, Benjamin, Meyers and Merrill [33] claimed that a large number of experimental results on end-linked polydimethylsiloxane networks are consistent with a modulus contribution from entanglements.

In contrast to most of the results from end-linked networks, very large entanglement contributions have been found in rubber networks made by cross-linking of long linear chains of polymers with large rubber plateau moduli. Moore and Watson [34] used peroxide cross-linking of natural rubber while high energy radiation was used by Dossin and Graessley [35] to make polybutadiene networks and by Pearson and Graessley [36] to make ethylene-propylene copolymer networks. The analysis used by Graessley and coworkers was based on the Langley theory [37]. The entanglement contribution to the elastic modulus was found to be several times larger than the contribution from chemical cross-links in both cases, and to depend on degree of cross-linking in the manner predicted by Langley. However, it is important to note that the entanglement contribution was found to decrease rapidly with polymer concentration during cross-linking, the entanglement contribution being roughly proportional to the volume fraction of polymer squared [36].

Ferry and coworkers also used high energy radiation for the cross-linking of 1,2-polybutadiene but the elastic contributions from chemical cross-links and entanglements, respectively, were separated experimentally by performing the cross-linking operation in the strained state.

Ferry's Two-Network Method

The principle of Ferry's Two-Network Method [38] is illustrated in Figure 5 for the case of simple extension. The reference state for the original uncross-linked polymer is the isotropic state where the sample has a length of $l_{ref;1}$. The sample is stretched to a length $l_{ref;2}$ at a temperature a few degrees above T_g and allowed to relax well into the rubber plateau. Before disentanglement of the chains can take place, the sample is quenched to a temperature well below T_g and cross-links are introduced in the glassy state. The cross-link network therefore has $l_{ref;2}$ as the reference length. After heating and release the sample retracts to the so-called state-of-ease in which the entanglement network and the cross-link network oppose each other with equal forces, i.e. the two effects have been separated experimentally. Equilibrium stress-strain

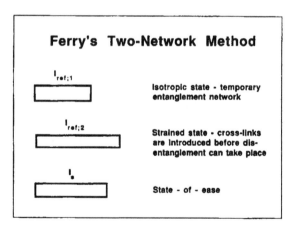

Figure 5. Principle of Ferry's Two-Network Method. An uncross-linked linear polymer is strained at a temperature slightly above T_g, quenched to the glassy state and cross-linked with high energy electrons. The sample retracts to the state-of-ease with length l, after heating and release (Reproduced from ref. 39 with permission from American Chemical Society)

measurements are made relative to the state-of-ease, allowing calculation of the elastic modulus in the state-of-ease.

Ferry initially based his method on the composite network theory of Flory [40]. This theory treats a network of Gaussian chains for which one set of cross-links is introduced in the isotropic state and a second set is introduced in the strained state. For such a seemingly complicated structure, Flory demonstrated for affine motion of the network junctions that the elastic free energy separates into two independent contributions, one from each of the two networks as shown in eq. 3.

$$\Delta A_{el} = \tfrac{1}{2} v_1 kT (\lambda_x^2 + \lambda_y^2 + \lambda_z^2 - 3) + \tfrac{1}{2} v_2 kT (\lambda_{x;2}^2 + \lambda_{y;2}^2 + \lambda_{z;2}^2 - 3) \qquad (3)$$

where v_1 and v_2 are the numbers of network strands per unit volume created in the isotropic and strained states, respectively; k is Boltzmann's constant; T is absolute temperature; λ_x, λ_y and λ_z are extension ratios relative to the reference state for the creation of the first set of network strands (the isotropic state); and $\lambda_{x;2}$, $\lambda_{y;2}$ and $\lambda_{z;2}$ are the extension ratios relative to the reference state for the creation of the second set of network strands. Flory's original expression also contained a volume term but this term has been omitted in eq. 3 since the volume is constant in the two-network experiment. Equation 3 substantiates the Two-Network Hypothesis of Andrews, Tobolsky and Hanson [41]. All the relevant elastic properties may be calculated from eq. 3.

Ferry realized that eq. 3 could be used for calculating the entanglement contribution of polymers cross-linked in the strained state by substituting the entanglement network for the network created in the isotropic state [38]. Results obtained at our own laboratory for three different types of strain are shown in Figure 6. It can be seen that the entanglement contribution to the shear modulus, G_N, is independent of type of deformation and approximately equal to the rubber plateau modulus G_N^o of a polybutadiene with a similar microstructure [1]. It can also be seen that G_N is much larger than the contribution from chemical cross-links, G_X. Extrapolation to zero strain during cross-linking gives G_N/G_X equal to three which means that entanglements contribute 75% of the equilibrium modulus in this case. However, the ratio decreases with increasing strain during cross-linking, indicating that eq. 3 is inadequate for a complete description of the properties at larger strains. Equation 3 also predicts isotropic properties relative to the state-of-ease in disagreement with the observed anisotropy for both elasticity and swelling [38,42,43].

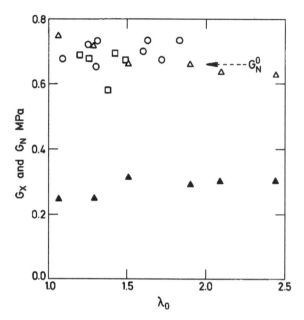

Figure 6. Elastic contributions from chain entangling G_N and chemical cross-links G_X in 1,2-polybutadiene at 50 °C plotted against extension ratio during cross-linking. Results of Hvidt, Kramer, Batsberg and Ferry [11] calculated from Flory's Composite Network Theory. Open symbols, G_N for three different types of strain. Solid triangles, G_X for simple extension. G_N^0 is the rubber plateau modulus of a polybutadiene with a similar microstructure (Reproduced with permission from American Chemical Society)

The strength of Ferry's Two-Network Method is that the two modulus contributions are compared directly without the need to know neither the concentration of network strands terminated by chemical cross-links nor the functionality or the functionality distribution of the cross-links.

Other molecular as well as purely empirical theories have been used for the analysis of the experimental data from Ferry's Two-Network Method [44-49]. The results of these calculations indicate additionally that the deviations from ideal stress-strain behavior originate in the entanglement contribution. Even so, it seems that the results of Ferry's Two-Network Method have not been generally accepted. Attempts have therefore been made at our own laboratory to simplify the analysis as well as the experiment itself in an effort to demonstrate that the main conclusions are of a general nature. The next section describes a model independent two-network calculation based on Hooke's Law. A two-network stress relaxation experiment, which requires no assumptions and no theory, is described in the last section.

Application of Hooke's Law

Hooke's Law applies to all solids and states that stress is proportional to strain at sufficiently small strains, Young's modulus being the coefficient of proportionality. For a composite rubber network which contains two different sets of cross-links, Hooke's Law may be applied to each of the two networks in the tradition of the work of Andrews, Tobolsky and Hanson [41]; Berry, Scanlan and Watson [50]; Flory [40]; Green, Smith and Ciferri [51]; and Ferry and coworkers [25,38]. The expression for the resulting unidirectional force f is shown in eq. 4 [39]

$$f = A_1 E_1 \frac{(l-l_{ref;1})}{l_{ref;1}} + A_2 E_2 \frac{(l-l_{ref;2})}{l_{ref;2}} \tag{4}$$

where A is cross-sectional area, l is sample length and E is Young's modulus. Subscript 1 refers to the first network which is formed with a reference length of $l_{ref;1}$ and subscript 2 refers to the second network which is formed with a reference length of $l_{ref;2}$. Equation 4 allows calculation of all the relevant elastic properties. The sample length l_s in the state-of-ease is calculated by setting the resulting force f equal to zero. The result is shown in eq. 5

$$\frac{E_2}{E_1} = \frac{l_{ref;2} \, (l_s - l_{ref;1})}{l_{ref;1} \, (l_{ref;2} - l_s)} \tag{5}$$

The modulus in the state-of-ease, E_s, is obtained by differentiation of eq. 4 with respect to l/l_s, the result being

$$E_s = E_1 + E_2 \tag{6}$$

Data by Hvidt, Batsberg, Kramer and Ferry [11] on 1,2-polybutadiene cross-linked in states of simple extension was used for the Hookean analysis. Young's modulus in the state-of-ease was obtained by fitting the original force-deformation data to a straight line at small strains. E_1 and E_2 were then calculated from eqs. 5 and 6. The results are summarized in Figure 7. Remembering that E=3G, comparison to Figure 6 shows that nearly identical results are obtained from Flory's Composite Network theory and from the Hookean analysis. The extrapolated value for the entanglement contribution to Young's modulus is $E_1=2.1$ MPa which is roughly equal to the rubber plateau

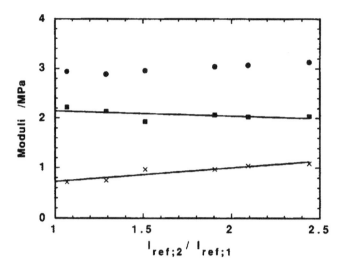

Figure 7. Young's moduli at 50 °C for 1,2-polybutadiene cross-linked in states of simple extension plotted against extension ratio during cross-linking: solic circles, E_s; solid squares E_1; crosses, E_2. Results of Twardowski and Kramer calculated from Hookean analysis [39] using original force-deformation data of Hvidt, Kramer, Batsberg and Ferry [11]. The entangle ment modulus obtained by linear extrapolation to zero strain is 2.14 MPa; the corresponding value for the chemical cross-links is 0.73 MPa (Reproduced with permission from American Chemical Society)

modulus of a 1,2-polybutadiene with a similar microstructure, $E_N^o = 2.0$ MPa [1]. The ratio of the modulus contributions from entanglements and chemical cross-links is again equal to three when extrapolated to zero strain during cross-linking.

The significance of the Hookean two-network analysis is that it is model and theory independent. The main results obtained by Ferry's Two-Network Method are therefore of a general nature.

Two-network stress relaxation experiment

A two-network experiment which requires no assumptions and no theory can be performed by a slight modification of Ferry's original two-network experiment. After the initial stretching to a length of $l_{ref;2}$ in Figure 5, the length is kept constant throughout the remainder of the experiment. Referring to Figure 8, this allows a direct comparison of the force f before quenching to the glassy state, the force f_c immediately after introduction of chemical cross

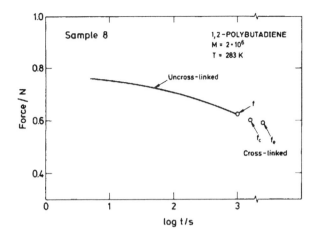

Figure 8. Two-network stress relaxation experiment. The forces are measured at constant length and temperature: f, immediately before quenching to the glassy state; f_c, after cross-linking and heating; and f_e, at elastic equilibrium (Results of Batsberg, Hvidt, Kramer and Fetters [53] reproduced with permission from Elsevier Science Publishers)

links by high energy radiation, and the force f_e at elastic equilibrium, all forces being measured both at the same length and the same temperature [52,53].

    To ensure an efficient trapping of the entangled structure at low degrees of cross-linking, Batsberg, Hvidt, Kramer and Fetters [53] used 1,2-polybutadiene with a very large molecular weight, $M=2x10^6$. Cross-linking was performed in the glassy state with 10 MeV electrons at doses of 50, 100 and 200 kGray (5, 10 and 20 Mrad). By comparison to the work of Hvidt, Batsberg, Kramer and Ferry [11], this should correspond to roughly 150, 300 and 600 cross-linked points per chain, respectively.

Table 1. Two-network stress relaxation results on 1,2-polybutadiene with a molecular weight of $2x10^6$. (Data from ref. 53)

	Sample 7	Sample 9	Sample 8
Dose (kGy)	50	100	200
100 f_c/f (%)	97	97	96
100 f_e/f (%)	91	92	95

The results are summarized in Table 1. The second row shows that the force measured immediately after the introduction of cross-links is 96-97 % of the force measured before cross-linking, independent of the degree of cross-linking. This means that the chemical cross-links contribute nothing to the measured force. The explanation is that the cross-links are introduced at the same length at which the force is measured.

The third row shows that the equilibrium force varies between 91 and 95 % of the force measured before cross-linking. Chain scission can therefore be disregarded. The slight dependence on degree of cross-linking indicates that the entangled structure in spite of the large numbers of cross-linked points per chain is incompletely trapped. The fact that f and f_e are nearly equal in a well cross-linked network demonstrates directly, that the contribution of trapped entanglements at elastic equilibrium and the stress measured on the uncross-linked polymer in the plateau region are equal. That this result is obtained without the need of any assumptions and even without the need of a theory should be emphasized.

The work discussed in the preceeding sections shows that G_N of eq. 2 is important and often much larger than G_X when undiluted or slightly diluted polymers of high molecular weight and with large rubber plateau moduli are cross-linked. In contrast to this, it has been concluded that G_N is zero or very small for most networks produced by end-linking [24,27]. For a certain polymer, both methods should give the same result unless one or both types of networks are less perfect than expected. We have begun work on end-linked systems in an effort to make end-linked networks with fewer defects [54] but progress has been slow.

Summary and Conclusion

It is demonstrated that the elastic and viscoelastic properties of elastomeric polymer networks are dominated by the effects of trapped and untrapped entanglements, respectively, at least for polymers with large values of the rubber plateau modulus.

Large mechanical losses are observed in lightly cross-linked rubber networks. The rate of relaxation of untrapped entanglements seems to depend exponentially on molecular weight of the dangling chain ends. This explains the exceedingly long relaxation times observed in the early stages of cross-linking of long linear chains. In contrast to this, short dangling chain ends relax very fast which explains why end-linked networks reach elastic equilibrium so quickly, even in the presence of substantial fractions of dangling chain ends.

Well cross-linked elastomers exhibit very small mechanical losses since the topological structure of chain entanglements is permanently trapped by the cross-links. The entanglement contribution to the equilibrium modulus at small strains is roughly equal to the rubber plateau modulus when the entangled structure has been efficiently trapped by cross-links.

It is concluded that chain entangling is a physical characteristic of flexible polymer chains, the elastic effect of which does not disappear on cross-linking but is strongly dependent on dilution during formation of the network. Theories of rubber elasticity which disregard elastic contributions from chain entangling other than a reduction of junction fluctuations must therefore be considered unphysical.

Acknowledgement

Financial support from the Danish Natural Science Research Council - Grant No. 5058 - and from the Danish Technical Research Council - Grant No. 5.17.1.6.09 is gratefully acknowledged.

References

[1] J.D. Ferry, Viscoelastic Properties of Polymers, 3rd ed., Wiley and Sons, New York, p. 374 (1980)
[2] Ref. 1, p. 383
[3] S. Onogi, T. Masuda, and K. Kitagawa, Macromolecules, 3, 109 (1970)
[4] O. Kramer, Brit. Polym. J., 17, 129 (1985)
[5] P.J. Flory, Proc. R. Soc. London. Ser. A., 351, 351 (1976)
[6] Ref. 1, p. 372
[7] P.G. deGennes, J. Chem. Phys., 55, 572 (1971)
[8] P.G. deGennes, Scaling Concepts in Polymer Physics, Cornell Univ. Press, Ithaca (1979)
[9] M. Doi and S.F. Edwards, J. Chem. Soc. Faraday Trans. II, 74, 1789 (1978)
[10] M. Doi and S.F. Edwards, The Theory of Polymer Dynamics, Clarendon Press, Oxford (1986)
[11] S. Hvidt, O. Kramer, W. Batsberg, and J.D.Ferry, Macromolecules, 13, 933 (1980)
[12] R.H. Valentine, J.D. Ferry, T. Homma, and K. Ninomiya, J. Polym. Sci. A-2, 6, 479 (1968)
[13] R. Chasset and P. Thirion, in: Proc. Conf. on Physics of Non-Crystalline Solids, J.A. Prins (Ed.), North Holland, Amsterdam, p. 345 (1965)
[14] O. Kramer, R. Greco, R.A. Neira, and J.D. Ferry, J. Polym. Sci., Polym. Phys. Ed., 12, 2361 (1974)
[15] O. Kramer, R. Greco, J.D. Ferry, and E.T. McDonel, J. Polym. Sci., Polym. Phys. Ed., 13, 1675 (1975)
[16] S. Granick, S. Pedersen, G.W. Nelb, J.D. Ferry, and C.W. Macosko, J. Polym. Sci., Polym. Phys. Ed., 19, 1745 (1981)

[17] G.W. Nelb., S. Pedersen, C.R. Taylor, and J.D. Ferry, J. Polym. Sci., Polym. Phys. Ed., 18, 645 (1980)
[18] H.-C. Kan, J.D. Ferry, and L.J. Fetters, Macromolecules, 13, 1571 (1980)
[19] R.E. Cohen and N.W. Tschoegl, Intern. J. Polymeric Mater., 2, 205 (1973)
[20] G.W. Kamykowski, J.D. Ferry, and L.J. Fetters, J. Polym. Sci., Polym. Phys. Ed., 20, 2125 (1982)
[21] P.G. deGennes, J. Physique, 36, 1199 (1975)
[22] D.S. Pearson and E. Helfand, Macromolecules, 17, 888 (1984)
[23] J.G. Curro, D.S. Pearson, and E. Helfand, Macromolecules, 18, 1157 (1985)
[24] P.J. Flory, Polymer, 20, 1317 (1979)
[25] J.D. Ferry, Polymer, 20, 1343 (1979)
[26] O. Kramer, Polymer, 20, 1336 (1979)
[27] J.E. Mark, Acc. Chem. Res., 18, 202 (1985)
[28] G. Heinrich, E. Straube, and G. Helmis, Adv. Polym. Sci., 85, 33 (1988)
[29] G. Ronca and G. Allegra, J. Chem. Phys., 63, 4990 (1975)
[30] P.J. Flory, J. Chem. Phys., 66, 5720 (1977)
[31] P.J. Flory and B. Erman, Macromolecules, 15, 800 (1982)
[32] B. Erman and P.J. Flory, Macromolecules, 15, 806 (1982)
[33] M. Gottlieb, C.W. Macosko, G.S. Benjamin, K.O. Meyers, and E.W. Merrill, Macromolecules, 14, 1039 (1981)
[34] C.G. Moore and W.F. Watson, J. Polym. Sci., 19, 237 (1956)
[35] L.M. Dossin and W.W. Graessley, Macromolecules, 12, 123 (1979)
[36] D.S. Pearson and W.W. Graessley, Macromolecules, 13, 1001 (1980)
[37] N.R. Langley, Macromolecules, 1, 348 (1968)
[38] O. Kramer, R.L. Carpenter, V. Ty, and J.D. Ferry, Macromolecules, 7, 79 (1974)
[39] T. Twardowski and O. Kramer, Macromolecules, in print
[40] P.J. Flory, Trans. Faraday Soc., 56, 722 (1960)
[41] R.D. Andrews, A.V. Tobolsky, and E.E. Hanson, J. Appl. Phys., 17, 352 (1946)
[42] R.L. Carpenter, H.-C. Kan, and J.D. Ferry, J. Polym. Sci., Polym. Phys. Ed., 18, 165 (1980)
[43] W. Batsberg, S. Hvidt, and O. Kramer, J. Polym. Sci., Polym. Lett. Ed., 20, 341 (1982)
[44] O. Kramer and J.D. Ferry, Macromolecules, 8, 87 (1975)
[45] R.L. Carpenter, O. Kramer, and J.D. Ferry, Macromolecules, 10, 117 (1977)
[46] R.L. Carpenter, O. Kramer, and J.D. Ferry, J. Appl. Polym. Sci., 22, 335 (1978)
[47] R.J. Gaylord, T.E. Twardowski, and J.F. Douglas, Polym. Bull., 20, 305 (1988)
[48] T.E. Twardowski and R.J. Gaylord, Polym. Bull., 21, 393 (1989)
[49] S. Hvidt, in: Physical Networks Polymers and Gels, W. Burchard and S.B. Ross-Murphy (Eds.), Elsevier Applied Science, London, p. 125 (1990)
[50] J.P. Berry, J. Scanlan, and W.F. Watson, Trans. Faraday Soc., 52, 1137 (1956)
[51] A. Green, K.J. Smith, and A. Ciferri, Trans. Faraday Soc., 61, 2772 (1965)
[52] W. Batsberg and O. Kramer, Rubber Chem. Technol., 54, 62 (1982)
[53] W. Batsberg, S. Hvidt, O. Kramer, and L.J. Fetters, in: Biological and Synthetic Polymer Networks, O. Kramer (Ed.), Elsevier Applied Science, London, p. 509 (1988)
[54] O. Kramer, in: Elastomeric Polymer Networks. A Memorial to Eugene Guth, J.E. Mark (Ed.), Prentice Hall, in print

Polymer Networks '91 pp. 79-98
Dosek and Kuchanov (Eds)
© VSP 1992

Structure and swelling properties of polymer networks synthesized in solution

Ferenc Horkay[1], Erik Geissler[3], Anne-Marie Hecht[3] and Miklos Zrinyi[2]

[1]*Institut für Makromolekulare Chemie, Universität Freiburg, Stefan Meier Strasse 31, D-7800 Freiburg i. Brsg. FRG*
[2]*Department of Colloid Science, Eötvös Loránd University, Pf. 32, H-1518 Budapest, Hungary*
[3]*Laboratoire de Spectrométrie Physique (CNRS associate Lab.), Université Joseph Fourier de Grenoble, B.P. 87, F-38402 St Martin d'Hères, France*

Abstract: The mechanical and osmotic properties of chemically cross-linked poly(vinyl acetate), polydimethylsiloxane and polyacrylamide gels have been investigated. Three types of measurements have been performed: (1) isotropic deswelling by means of lowering the activity of the diluent; (2) anisotropic deswelling induced by uniaxial compressional stress; (3) measurement of the shear modulus at different polymer concentrations. In addition small angle X-ray scattering measurements were performed to reveal the structural differences between the gel and the uncross-linked polymer solution. The results are consistent with a free energy function containing separable elastic and mixing contributions. The mixing term for the network polymer differs from that of the corresponding polymer solution and is independent of the solvent quality.

INTRODUCTION

One of the basic hypotheses of theories [1-6] describing the swelling behaviour of polymer gels is that the elastic and the mixing contributions in the free energy that accompanies the swelling of the dry network are separable, i.e.

$$\Delta F = \Delta F_{el} + \Delta F_{mix} \qquad\qquad (1)$$

A second assumption is that ΔF_{mix} for the cross-linked polymer is identical with the free energy of mixing for the solution of the uncross-linked polymer of infinite molecular weight. This assumption implicitly presumes that the polymer-solvent interaction parameter is not affected by the presence of cross-links.

Two further assumptions concern the elastic free energy term: this term is supposed to be independent, firstly, of the solvent quality, and secondly, of the symmetry of deformation. This latter assumption means that no distinction is made between the free energy of mixing for an isotropically swollen gel and that for the same gel swollen to the same extent under an external anisotropic constraint.

To test the validity of these assumptions simultaneously is rather difficult because the result strongly depends on the particular theoretical model used to evaluate the experiments. It is usually supposed that the elastic term can be described by one of the existing network models (affine or non-affine models). Mechanical measurements allow the elastic free energy to be computed for the swollen network. Reanalysis of previously published data by Gottlieb and Gaylord [7] led to the conclusion that none of the molecular models of network elasticity fits satisfactorily the experiments, i.e. there is no objective criterion to select a specific molecular model. An alternative approach to the problem is the measurement of solvent activity differences between cross-linked and uncross-linked polymers [8]. High precision vapour pressure measurements of Eichinger and co-workers [9-12] have indicated deviation from the additivity of elastic and mixing free energies. The observed nonadditivity has been attributed to possible solvent dependence of the elastic term.

Because of the unsatisfactory understanding of the
molecular origin of network elasticity and the
uncertainty concerning cross-link dependence of the
mixing term, simultaneous measurement of both
mechanical and thermodynamic properties of swollen
network systems allows a more complex test of the
validity of assumptions involved in the theory.

In the present work swelling pressure and the shear
modulus measurements performed on chemically different
gel systems are reported, together with the osmotic
pressure data of the corresponding polymer solutions.
These measurements allow a model-independent comparison
to be made between the gel and the solution mixing term.
Comparison is also made between elastic modulus data
obtained from isotropic swelling pressure measurements
and from uniaxial compression measurements to check the
separability of the elastic contribution in the total
free energy function of the gel. The effect of solvent
quality on the elastic free energy as well as that of an
anisotropic external constraint on the mixing
contribution is investigated.

EXPERIMENTAL PART

Gel Preparation

Poly(vinyl acetate) (PVAc) gels were prepared by
acetylation of chemically cross-linked poly(vinyl
alcohol) networks using a method described elsewhere
[13-14]. Cross-links were introduced into aqueous
poly(vinyl alcohol) solutions of concentration 3,6,9
and 12% w/w. The cross-linking agent was glutaraldehyde.
At each concentration several gel samples were prepared
with different molar ratios of monomer units to the
molecules of the cross-linking agent. The gels are
designated by the code X/Y, where X is the concentration
of the polymer at cross-linking in percentage weight

fraction, and Y is the ratio of monomer units to cross-
linker. The acetylation of the poly(vinyl alcohol) gels
was performed in a mixture of pyridine-acetic acid-
acetic anhydride at 90 °C. After completion of the
reaction the gels were washed and then swollen to
equilibrium in good solvents (toluene and acetone at
25 °C) and in a theta solvent (isopropyl alcohol at
52 °C), respectively.

Polydimethylsiloxane (PDMS) gels were prepared [15-16]
by end-linking of hydroxyl terminated PDMS chains of
approximate molecular weight 40.000 with the cross-
linking agent ethyl triacetoxy silane. Filled PDMS gel
samples containing 10% by weight of fumed silica beads
were also made. The samples without silica were prepared
in chlorobenzene at 40% and 60% by volume of the polymer.
The filled gels were made from the undiluted polymer.
After completion of the reaction, the sol fraction was
extracted by octane. The swelling and mechanical
properties of the PDMS gels were studied in toluene at
25 °C (good solvent).

Poly(acrylamide-bisacrylamide) hydrogels were prepared
[17] at room temperature (23 °C) at acrylamide
concentration 0.08 gcm^{-3}. The concentration of N,N'-
methylenebisacryamide was varied in the range 0.001 to
0.005 w/w. The gelation was initiated with 0.7 g/L
ammonium persulfate and 5.7×10^{-4} v/v tetramethylene-
ethylenediamine. The gels were allowed to swell in
distilled water at 25 °C.

Methods

The swelling pressure and the shear modulus of the gels
were determined as a function of the concentration.
Deswelling was induced in the swollen networks by
surrounding them with polymer solutions of known osmotic
pressure (Refs.18-19). The PVAc and the PDMS gels were

equilibrated with PVAc-toluene solutions, while in the case of the polyacrylamide hydrogels the water activity was adjusted by poly(vinyl pyrrolidone). The gels were enclosed in dialysis bags to prevent penetration of the polymer molecules [20].

Shear modulus, G_s, data were obtained from uniaxial compression measurements performed on cylindrical gel samples at constant volume, in an apparatus described elsewhere [21]. The stress-strain data in the range of deformation ratio $0.7 \leq \Lambda < 1$, were evaluated using the Mooney-Rivlin equation. The C_2 term was found to be equal to zero.

Small angle X-ray scattering (SAXS) measurements [22] were performed on PVAc/acetone gels and solutions using the D24 instrument at the Laboratoire pour l'Utilisation du Rayonnement Electromagnétique (LURE), Orsay, France. Samples were placed between two mica sheets separated by a 1 mm annular spacer. A 6 cm linear detector with 256-point resolution was used, with a sample-detector distance of 1.2 m. The incident wavelength was 1.608 Å. An acetone sample in the same cell was used for background substraction, and the resulting difference signal was normalised by the independently measured detector response.

RESULTS AND DISCUSSION

The swelling pressure of the gel, ω, which is a directly measurable thermodynamic quantity, is the derivative of the total free energy with respect to the polymer volume fraction, φ.

$$\omega = \varphi^2 \frac{\partial(\Delta F / \varphi)}{\partial \varphi} \qquad (2)$$

According to the principle of additivity of the elastic and mixing contributions we have

$$\omega = \Pi_m - G_v \qquad (3)$$

where G_v is the elastic modulus of the gel and Π_m is the mixing pressure of the network polymer.

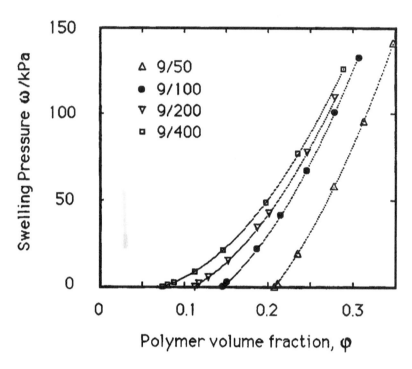

Fig.1. Swelling pressure, ω, as a function of polymer volume fraction, φ, for PVAc/toluene gels. The curves are the least-squares fits to Eq.(4), with parameters given in Table 1.

In Fig.1 a typical set of swelling pressure *vs.* polymer volume fraction data are shown for PVAc gels swollen in toluene (Ref.23). The gels differ in the extent of cross-linking. It can be seen that the swelling pressure rapidly increases with decreasing the swelling degree. The swelling pressure curves were analysed using a two term equation

$$\omega = A\varphi^n - G_o\varphi^m \qquad (4)$$

where the first term represents the mixing and the

second term the elastic contribution. In principle Eq.(4) has been proposed by Flory [24], who applied a similar analysis to estimate the molecular parameters of swollen rubber networks from their equilibrium concentration and elastic modulus data.

Table 1
Fitting parameters of Eq.(4) to swelling pressure data for chemically different network systems

Sample	φ_e	G_o/kPa	n	m
PVAc/toluene (good solvent condition)				
6/50	0.146	59.7	2.29	0.331
6/200	0.078	17.1	2.22	0.340
9/50	0.208	123.6	2.27	0.355
9/200	0.112	33.9	2.27	0.326
12/50	0.229	168.3	2.35	0.383
12/200	0.133	50.1	2.26	0.335
PVAc/acetone (good solvent condition)				
9/100	0.103	51.6	2.24	0.321
9/200	0.078	36.0	2.26	0.346
9/400	0.051	16.3	2.24	0.369
PVAc/isopropyl alcohol (theta condition)				
6/50	0.253	66.0	2.78	0.326
6/200	0.149	26.6	2.61	0.413
9/50	0.330	95.2	3.70	0.252
9/100	0.237	53.5	3.00	0.331
9/200	0.201	37.5	2.85	0.321
PDMS/toluene (good solvent condition)				
unfilled 40	0.068	18.4	2.22	0.347
unfilled 60	0.095	35.2	2.21	0.343
filled (10% silica)	0.202	146.1	2.29	0.327
Polyacrylamide/water				
8/1	0.036	14.1	2.61	0.38
8/2	0.048	12.8	2.82	0.26
8/3	0.055	19.0	2.96	0.33
8/4	0.065	29.2	2.98	0.40
8/5	0.066	21.7	3.10	0.35

In Eq.(4) the constant A depends on the particular polymer-solvent system, and the exponent n equals to 3 or 9/4 according to the classical or the scaling

theory [25], respectively. G_0 is proportional to the
concentration of the network chains. For the exponent m,
rubber elasticity theory [26] predicts the value 1/3.
In Table 1 the results of the least squares fit of
Eq.(4) to the experimental data are shown for chemically
different network systems (φ_e is the concentration of the
freely swollen gel). In good solvent condition the
exponents are close to the values of 9/4 and 1/3,
respectively. In the case of the theta system [27] the
experimentally observed n scatters around 3. In Fig.1 it
is apparent that the calculated curves describe
satisfactorily the experimental values of the swelling
pressure over the whole concentration range.
It is interesting to compare the behaviour of the
unfilled PDMS gels with those containing silica filler
particles [16]. Neutral filler particles are not expected
to make a significant contribution to the osmotic
properties of the swollen network. It is reasonable to
assume that they make their influence felt by occupying
a fixed volume in the system, consequently the osmotic
contribution to the swelling pressure is governed by the
network polymer alone. Performing the analysis on the
basis of this assumption the exponents n=2.29 and
m=0.327 are obtained, which are in good agreement with
those obtained for the unfilled PDMS gels.

Significant deviation from the theoretical exponents
can only be observed in the case of the polyacrylamide
gels as the cross-linking density increases [28]. It is
known that in polyacrylamide networks during the
gelation structural heterogeneities develop having
characteristic size in the range of 50-2500 Å [29].
These heterogeneities are expected to influence both
the osmotic and the elastic properties of the swollen
networks. The description of the behaviour of such non-
homogeneous systems is beyond the scope of existing

molecular theories of network elasticity. It is however
interesting to note that the concentration dependence
of the elastic modulus exhibits no significant deviation
from the theoretical prediction even for the most densily
cross-linked polyacrylamide gels.

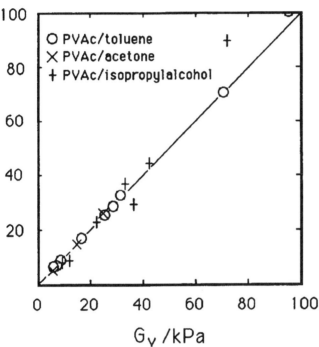

Fig.2. Plot of the shear modulus, G_s, as a function of
the osmotic modulus, $G_v (=G_0 \varphi_e^m)$, for PVAc/toluene,
PVAc/acetone and PVAc/isopropyl alcohol gels.

A crucial test of the separability of the elastic and
mixing free energies is to compare the elastic modulus
derived from the swelling pressure according to Eq.(4)
with the directly measured shear modulus, G_s. In Fig.2
G_s is presented as a function of the osmotic modulus
$G_v (=G_0 \varphi_e^m)$ calculated by parameters given in Table 1.

It is apparent that the numerical values of the shear
and the osmotic moduli are identical within the limit
of experimental error [30].

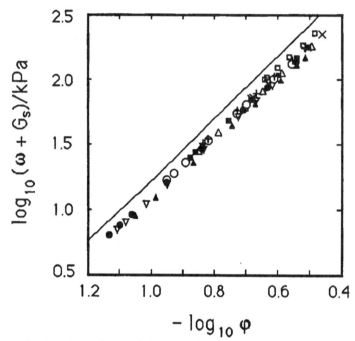

Fig.3. Double logarithmic plot of $(\omega+G_s)$ *vs* φ for
PVAc/toluene gels. Continuous line represents
Π_m for the PVAc solution $(M=\infty)$. Different symbols
refer to different gel samples.

In order to estimate the mixing pressure of the cross-
linked polymer in Fig.3 we plot $\omega+G_s$ as a function of
the polymer volume fraction for PVAc/toluene gels. The
continuous line shows the concentration dependence of
the osmotic pressure of the uncross-linked polymer of
infinite molecular weight. If the osmotic contribution
of the gel and the osmotic pressure of the solution were
equal the two curves would coincide. In the double
logarithmic plot it can be seen that $\omega+G_s$ is considerably
smaller than Π_m, but both quantities follow similar
dependence on the concentration. Agreement between the

two curves can be achieved if the polymer concentration
in the gel is decreased by a factor of 0.86 for
PVAc/toluene gels and 0.80 for PVAc/acetone gels [23].

Mixing Pressure
Π_m, $\omega + G_s$
   /kPa

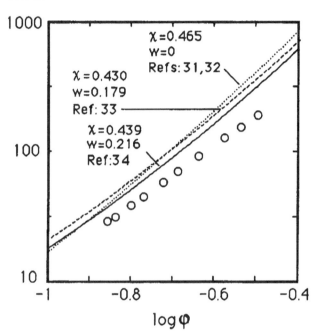

Fig.4. Double logarithmic plot of $(\omega + G_s)$ *vs* φ for
      PDMS/toluene gels. The curves represent Π_m for
      the PDMS solution (M=∞) calculated from data
      reported in the literature [31-34].

In Fig.4 $\omega + G_s$ is shown as a function of the
concentration for a PDMS/toluene gel together with the
mixing pressure, Π_m, for the PDMS solution. In the
figure, Π_m *vs* φ curves calculated from data reported by
different authors [31-34] are also shown. These results,
obtained by entirely different techniques (swelling
measurements, static light scattering and small angle
neutron scattering measurements) are in satisfactory

agreement. The solution curves always lie above the gel
curve. In Fig.4 are also displayed the parameters from
the fit of the data of different authors to the Flory-
Huggins equation

$$\Pi_m = -(RT/v_1) \, [\ln(1-\varphi)+\varphi+\chi_1\varphi^2+w\varphi^3] \tag{5}$$

where χ_1 and w are interaction parameters.

These observations can be interpreted in terms of a
strong decrease in the degree of freedom of the swollen
network due to the presence of cross-links. Inter-
connection of the chains reduces the number of
configurations accessible to the cross-linked system.
The local segment density is expected to increase in
the vicinity of the junction points. In the swollen
network regions of excess polymer content build up that
appear as permanent departures from uniformity.
Consequently the effective polymer concentration that
controls the thermodynamic behaviour of the swollen
network should be smaller than the average concentration.
According to this picture the polymer-polymer
correlation length must be longer and the osmotic
contribution must be smaller in the gel than in the
corresponding polymer solution. Such a change in the
correlation length should be directly observable in
small angle scattering measurements.

The scattering intensity I(Q) of a polymer solution
is directly related to the osmotic compressibility

$$I(Q) = a \, \frac{kT(\rho_p-\rho_s)^2\varphi^2}{K_{os}} \, S_S(Q) \tag{6}$$

where *a* is a constant depending on scattering geometry
used, K_{os} ($=\varphi\partial\Pi/\partial\varphi$) is the osmotic compressional modulus,
ρ_p and ρ_s are the electron densities of the polymer and
the solvent respectively, $S_S(Q)$ is the structure factor
of the solution, and $Q=(4\pi/\lambda)\sin(\theta/2)$ is the transfer
wave vector for an incident wavelength λ and scattering

angle θ. For semi-dilute solutions, at small values of Q the structure factor can be approximated by an Ornstein-Zernicke lineshape function

$$S_S(Q) = 1/(1+Q^2\xi^2) \qquad (7)$$

where ξ is the density-density correlation length in the solution [25].

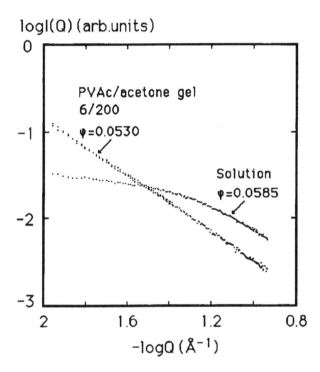

Fig.5. Small angle X-ray scattering spectra of PVAc/acetone gel 6/200 (φ_e=0.053) and PVAc/acetone solution at φ=0.0585.

In the swollen network regions of excess polymer content build up that appear as permanent, static departures from uniformity. The total amplitude of the concentration fluctuations is the sum of two contributions, a static and a dynamic part [15,35]:

$$\langle\Delta\varphi^2\rangle = \langle\Delta\varphi^2\rangle_{dyn} + \langle\delta\varphi^2\rangle \qquad (8)$$

It is reasonable to assume that the dynamic part, which
exhibits solution-like behaviour, is represented by a
structure factor of the same form as Eq.(7). The static
contribution, $<\delta\varphi^2>$, depends upon the size and
distribution of the non-uniformities, and hence on the
particular network system.

In Fig.5 the X-ray scattering spectrum of a PVAc/acetone
gel is shown together with that of the corresponding
solution [22]. Comparing the solution and the gel spectra
at almost the same concentration the difference between
the two curves is apparent. In the double logarithmic
representation the excess scattering from the gel at
small Q values indicates the presence of large static
scatterers. At increasing Q the discrepancy between the
gel and solution decreases and at about Q=0.08 $Å^{-1}$ the
two curves cross over so that at higher Q values the
scattering from the solution exceeds that of the gel.

In Fig.6 the spectrum of the same PVAc/acetone gel is
compared with that of a solution with decreased polymer
concentration. Now it can be seen that in the high Q
region of the spectra, where the static contribution to
the gel signal becomes negligible, coincidence is
obtained between the two spectra. This result is
consistent with the intuitive picture that the polymer
concentration is enhanced around the cross-links and
depleted in the rest of the sample, i.e. the elastic
constraints generated by the cross-links cause non-
uniformities in the polymer concentration. The static
non-uniformities cause excess scattering intensity in
the small Q region but do not contribute significantly
to the osmotic properties of the swollen network.

In order to check the dependence of the elastic free
energy on the quality of the solvent in Fig.7 the shear
moduli of PVAc gels measured in a theta solvent (iso-
propylalcohol at 52 °C) are compared with those obtained
for the same gels in a good solvent (toluene at 25 °C) at

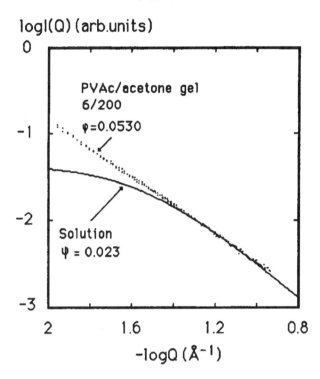

logI(Q) (arb.units)

PVAc/acetone gel
6/200
ψ=0.0530

Solution
ψ = 0.023

$-\log Q\ (\text{Å}^{-1})$

Fig.6. Spectrum of PVAc/acetone gel 6/200 (experimental
points). Continuous line: spectrum of solution
with φ adjusted so that the high Q region of the
spectrum match the gel.

identical concentration. After application of the
correction for absolute temperature the values of the
moduli in the good solvent are the same as in the theta
diluent. This result indicates that the elastic free
energy of the swollen gel is unaffected by the quality
of the solvent other than the swelling equilibrium
concentration. Similar results has been reported recently
by McKenna et al. [36-37] for rubber networks swollen in
different solvents.

A further assumption of the theory is that the free
energy of mixing is identical in the isotropically
swollen gel and in the same gel swollen to the same
volume under an anisotropic constraint.

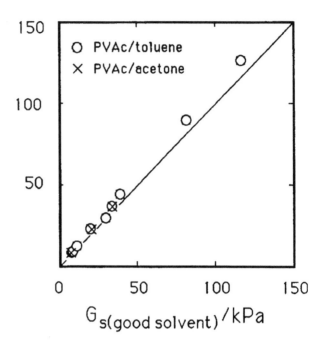

$\underset{\ominus}{\overset{I}{\ominus}}$ $G_{s(theta\ solvent)}$ /kPa

$G_{s(good\ solvent)}$/kPa

Fig.7. Plot of the shear modulus, G_s, of PVAc gels in
       theta solvent (isopropyl alcohol at 52 °C) as a
       function of the shear modulus measured in good
       solvent (toluene at 25 °C) at the same
       polymer concentration

The free energy of the swollen constraint network is
given as [26]

$$\Delta F = \Delta F_{mix} + (G_0/2)(\alpha_x^2 + 2/\varphi\ \alpha_x - 3) \qquad (9)$$

where α_x is the deformation ratio relative to the dry
network in the x direction. In equilibrium with a pure
diluent we have

$$\Pi_m - G_0\varphi^{1/3}/\Lambda = 0 \qquad (10)$$

where Λ is the uniaxial compression ratio of the deformed gel. Eq.(10) is valid only if the mixing contribution is independent of the symmetry of the deformation. If any anisotropic contribution to the mixing free energy is present then it is expected to increase with increasing Λ.

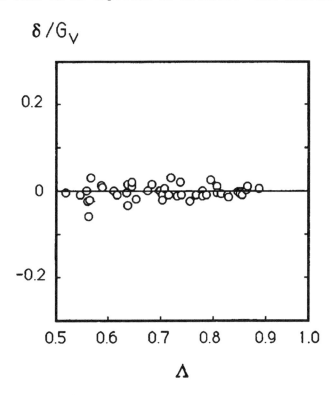

Fig.8. Plot of δ/G_v as a function of the deformation ratio, Λ, for PVAc/toluene gels. The quantity $\delta(=\Pi_m-G_o\varphi^{1/3}/\Lambda)$ represents the difference between the mixing pressures of the isotropically and the unisotropically swollen gel at identical polymer concentration.

In Fig.8 the difference between Π_m obtained from isotropic swelling pressure observations and the quantity $G_o\varphi^{1/3}/\Lambda$ calculated from uniaxial compression measurements is shown in the form of δ/G_v as a function of the deformation ratio, Λ. The quantity $\delta(=\Pi_m-G_o\varphi^{1/3}/\Lambda)$

represents the difference in mixing pressure between the isotropically and unisotropically swollen states at the same polymer concentration. No systematic trend can be detected with increasing the deformation [38-39].

CONCLUSIONS

Osmotic and mechanical properties of chemically cross-linked network systems have been investigated. The results show, that for lightly cross-linked networks:

- the total free energy that accompanies swelling is separable to an elastic and a mixing free energy term;

- the mixing term for the network polymer differs from the free energy of mixing of the uncross-linked polymer of infinite molecular weight;

- the elastic term is invariant with respect to the solvent quality;

- the mixing term is independent of the state of strain in the gel.

ACKNOWLEDGEMENTS
This work is part of a joint CNRS-Hungarian Academy of Sciences project. We also acknowledge research Contract OTKANo.2158 from the Hungarian Academy of Sciences. We are grateful to LURE, Orsay, for access to the D24 instrument on the DCI synchrotron. F.H. acknowledges a research fellowship from the Alexander von Humboldt Stiftung.

REFERENCES

[1] P.J.Flory,J.Rehner,J.Chem. Phys. 11, 521 (1943)

[2] H.M.James,E.Guth,J.Chem. Phys. 11, 455 (1943)

[3] J.Frenkel,Rubber Chem.Technol. 13, 264 (1940)

[4] P.J.Flory,Principles of Polymer Chemistry. Cornell, Ithaca, NY, (1953)

[5] P.J.Flory,J.Chem.Phys. 66, 5720 (1977)

[6] P.J.Flory,B.Erman,Macromolecules 15, 800 (1982)

[7] M.Gottlieb,R.G.Gaylord,Macromolecules 17, 20 (1984)

[8] G.Gee,J.B.M.Herbert,R.C.Roberts,Polymer 6, 54 (1965)

[9] R.W.Brotzman,B.E.Eichinger,Macromolecules 14, 1445 (1981)

[10] R.W.Brotzman,B.E.Eichinger,Macromolecules 15, 531 (1982)

[11] R.W.Brotzman,B.E.Eichinger,Macromolecules 16, 1131, (1983)

[12] N.A.Neuburger,B.E.Eichinger,Macromolecules 21, 3060 (1988)

[13] F.Horkay,M.Nagy,M.Zrinyi,Acta Chim.Acad.Sci.Hung. 108, 287 (1981)

[14] F.Horkay,M.Zrinyi,Macromolecules 15, 1306 (1982)

[15] S.Mallam,A.M.Hecht,E.Geissler,P.Pruvost,J.Chem. Phys. 91, 6447 (1989)

[16] F.Horkay,E.Geissler,A.M.Hecht,P.Pruvost,M.Zrinyi, Polymer 32, 835 (1991)

[17] S.Mallam,F.Horkay,A.M.Hecht,E.Geissler, Macromolecules 22, 3356 (1989)

[18] H.Vink,Europ. Polym. J. 7, 1411 (1971)

[19] H.Vink,Europ. Polym. J. 10, 149 (1974)

[20] M.Nagy,F.Horkay,Acta Chim.Acad.Sci.Hung. 104, 49 (1980)

[21] F.Horkay,M.Nagy,M.Zrinyi,Acta Chim.Acad.Sci.Hung. 103, 387 (1980)

[22] A.M.Hecht,F.Horkay,E.Geissler,J.P.Benoit, Macromolecules 24, 4183 (1991)

[23] F.Horkay,A.M.Hecht,E.Geissler,J. Chem. Phys. 91, 2706 (1989).

[24] P.J.Flory,Ind. Eng. Chem. 38, 417 (1946)

[25] P.G.de Gennes,Scaling Concepts in Polymer Physics. Cornell: Ithaca, NY, 1979.

[26] L.R.G.Treloar,The Physics of Rubber Elasticity. Clarendon, Oxford 3rd Ed. 1975.

[27] F.Horkay,E.Geissler,A.M.Hecht,M.Zrinyi, Macromolecules 21, 2589 (1988)

[28] F.Horkay,A.M.Hecht,E.Geissler,Macromolecules 22,
      2007 (1989)

[29] A.M.Hecht,R.Duplessix,E.Geissler,Macromolecules
      18, 2167 (1985)

[30] E.Geissler,F.Horkay,A.M.Hecht,M.Zrinyi,J.Chem.Phys.
      90, 1924 (1989)

[31] N.R.Langley,J.D.Ferry,Macromolecules 1, 353 (1968)

[32] H.Shih,P.J.Flory,Macromolecules 5, 761 (1972)

[33] V.K.Soni,R.S.Stein,Macromolecules 23, 5257 (1990)

[34].A.M.Hecht,A.Guillermo,F.Horkay,J.F.Legrand,
      E.Geissler, Macromolecules submitted

[35] F.Horkay,A.M.Hecht,S.Mallam,E.Geissler,A.R.Rennie,
      Macromolecules 24, 2896 (1991)

[36] G.B.McKenna,K.M.Flynn,Y.Chen,Polym. Commun. 29, 272
      (1988)

[37] G.B.McKenna,K.M.Flynn,Y.Chen,Macromolecules 22, 450
      (1989)

[38] F.Horkay,M.Zrinyi,Macromolecules 21, 3260 (1988)

[39] A.M.Hecht,F.Horkay,E.Geissler,M.Zrinyi,Polym.Commun
      31, 53 (1990)

Polymer Networks '91 pp. 99-118
Dosek and Kuchanov (Eds)
© VSP 1992

Thermoelasticity and strain-induced volume-effects in permanent networks

H.G. Kilian

Abteilung Experimentelle Physik, Universität Ulm, Oberer Eselsberg, 7900 Ulm, Germany

ABSTRACT

The fundamental equation of a network is defined by using the van der Waals approach. Thermoelastic phenomena are fairly well interpreted including that the thermoelastic inversion does not appear in "pressure-induced" modes of deformation. In analogy, the volume of a net-work is predicted to decrease in pressure-induced modes while the system expands in "stress-induced" modes. The strain dependence of internal energy is anti-symmetric for the same reason. These phenomena are described by using the thermodynamic coefficients as measured in the unstrained network. It is shown that a van der Waals network behaves as a weakly interacting conformational gas. Phenomena in the local regime are adjusted to the global level showing liquidlike properties up to largest strains. This explains, why the fundamental equation can be formulated as presented.

INTRODUCTION

It is still outstanding to know the Gibbs-function of a thermodynamic system [1]. Under the few systems with a defined Gibbs-function are gases. Collisions between particles guarantees equipartition of energy. In this respect, it is now an interesting matter that strained networks behave in the global regime like a conformational gas [2]. Kinetic energy is equipartitioned [3]. Global interaction between the chains should therefore have the same nature as in gases. Stochastic collisions [4] should happen to be the elementary process. In view these analogies it is clear that it should be possible to formulate the fundamental equation of networks [5].

Yet, comparing the ideal gas and the Gaussian network there is the difference that each phantom-chain itself is a statistic system [6,7] with liquid-like local properties. Every Gibbs-function of a network has therefore to define how global and local properties are interrelated [6,7,8,9,10,11].

Adequate experiments were done by using special equipments [6,7,10-15]. One of these devices is a stretching micro-calorimeter [16-20] that allows to measure the whole energy balance during deformation. This method goes back to A. Engelter and F.H.Müller. Another significant method is to measure the volume changes during uniaxial deformation [21].

In this paper we recall thermodynamics of van der Waals-networks. A treatment of different types of strain is included [2,22-24]. The fundamental equation of a network is formulated. Its reliability is then probed by discussing the strain-induced transformation of intrinsic properties in different modes of deformation.

THE DIFFERENTIAL OF THE FUNDAMENTAL EQUATION

In literature the Helmholtz free energy defined in the system of coordinates $\{T,V,L\}$ is often used [6,7]

$$dF = -SdT - pdV + K_a dL \qquad (1)$$

T is the absolute temperature, p is the pressure. L represents the single independent extensive deformation variable, the actual length of the deformed network, The formulation in equation (4) includes uniaxial and biaxial modes of deformation in the last case with a fixed ratio of both strains so that the single independent and extensive deformation variable L is left. S is the entropy, V is the volume. K_a is the force applied in the deformation mode α. Nominal force f_a and force K_a are interrelated according to

$$K_\bullet = A_o f_\bullet \tag{2}$$

where A_o is the initial cross-section. A severe problem is now that the most relevant Maxwell-relation

$$(\partial S/\partial L)_{v,T} = -(\partial K_\bullet/\partial T)_{v,L} \tag{3}$$

is formulated under constant length- and volume-conditions. Any experiment under these conditions is difficult [10]. It was undertaken to transform this equation directly into the system {T,p,L} [9].

But, it is much better to use a Legendre-transformation [1] for changing the system of coordinates. We are then led to the Gibbs free energy [20-,12,13] the differential dG of which is equal to

$$dG = -SdT + Vdp + K_\bullet dL \tag{4}$$

$$\text{(I)} \qquad \text{(II)}$$

It does not correspond to the usual treatment of deformation of an homogeneous elastic continuum [14] that strain-energies exchanged during compression (I) or during deformation under constant volume conditions (II) are defined as independent energy forms. The formulation implicates three Maxwell-relations

Temperature-Strain Relation (thermoelasticity)

$$[\partial S/\partial L]_{p,T} = -[\partial K_\bullet/\partial T]_{p,L} = A_o(\partial f_\bullet/\partial T)_{p,L} + f_\bullet(\partial A_o/\partial T)_{p,L} \tag{5}$$

We learn from this equation that elastic and thermal properties of the network should be uniquely coupled. With the aid of

$$dQ = TdS \tag{6}$$

one finds

$$\left[\partial Q/\partial L\right]_{p,T} = T\left[\partial K_{\bullet}/\partial T\right]_{p,L} \tag{7}$$

Hence, to measure the heat exchanged during deformation is a method for studying thermo-elastic phenomena directly. The results may then be described by deducing the temperature coefficient of the force from the mechanic equation of state.

Pressure-Strain Relation

$$\left(\partial V/\partial L\right)_{T,p} = \left(\partial K_{\bullet}/\partial p\right)_{T,L} \tag{8}$$

Strain-induced changes of the volume and the compressibility of the network are interrelated here. This relationship explains the necessary volume effects during deformation.

Temperature-Pressure Relation

$$\left[\partial S/\partial p\right]_{T,L} = \left[\partial V/\partial T\right]_{p,L} \tag{9}$$

In this relation thermal expansion is linked with the pressure induced change of entropy leading to a situation which is not easily probed by experiment.

The equations (5) and (8) are thus key-relations for studying whether condensed matter properties and global properties of networks are interrelated so as to justify the fundamental equation as in equation (4).

THERMO-ELASTIC SYMMETRIES

First, we want to elucidate an interesting symmetry. To this end, let us define the total force $K_{\bullet}$ as the sum of $K_{\bullet}^{(h)}$ as the enthalpy- and $K_{\bullet}^{(s)}$ as the entropy-component [6,7,5]

$$K_{e} = (\partial G / \partial L)_{p,T} = (\partial H / \partial L)_{p,T} - T (\partial S / \partial L)_{p,T} = K_{e}^{(h)} + K_{e}^{(s)} \qquad (10)$$

If we now write the force so that it is explicitly expressed that the modulus is proportional to the temperature

$$K_{e} = T \Xi_{e} (T,p) \qquad (11)$$

we find the entropy component of the force

$$K_{e}^{(s)} = T \partial K_{e} / \partial T = K_{e} + T (\partial \Xi_{e} / \partial T)_{L,p} \qquad (12)$$

comprised of two terms, the first one giving the global force exerted onto the network while the second one is addressed to strain-induced changes of intrinsic properties. According to equation (10) and equation (12) we have

$$K_{e}^{(h)} = K_{e} - K_{e}^{(s)} = -T (\partial \Xi / \partial T)_{L,p} \qquad (13)$$

what yields to the symmetric relationship

$$K_{e}^{(h)} = (\partial H^{intrins} / \partial L)_{T,p} = T (\partial \Xi / \partial T)_{L,p} = T (\partial S^{intrins} / \partial L)_{T,p} = K_{e} - K_{e}^{(s)} \qquad (14)$$

Annihilation or production of internal states leads to saturation yielding with equation (14) to the condition of internal equilibrium

$$(\partial G^{intrins} / \partial L)_{T,p} = 0 \qquad (15)$$

One of the most prominent examples is the strain-induced production or the annihilation of holes making a liquid a "hole-saturated system". In the local regime this should also hold true for networks because of having here liquidlike properties. Hence, it is due to equation (15) that "Internal equilibrium properties" have nearly no influence on the isothermal and isobaric stress-strain behaviour.

THE VAN DER WAALS-APPROACH

The strain-energy of a van der Waals-network in the deformation mode α is derived to be given by [18,19]

$$W_\alpha = -G_o\left[[\ln(1-\eta_\alpha)+\eta_\alpha]2\phi_{max}+(2/3)a\phi_\alpha^{3/2}\right]=G_oW_\alpha^* \tag{16}$$

with

$$\phi_\alpha = (I_{1,\alpha}-3)/2 \; ; \; \phi_{max}=(1/2)\left(\lambda_{max}^2+2/\lambda_{max}-3\right); \; I_{1,\alpha}=\sum_{i=1}^{3}\lambda_{i,\alpha}^2 \tag{17}$$

and

$$\eta_\alpha = \sqrt{\phi_\alpha/\phi_{max}} \tag{18}$$

The modulus is written as

$$G_o = (\rho RT/M_c)\left(<r^2>/<r_o^2>\right) \tag{19}$$

where ϱ is the density. M_c is the molecular weight of the chains. The last factor is the memory-term [15,6] which relates the mean square of the chain-end-to-end vectors $<r^2>$ to the unpertubated dimensions $<r_o^2>$.

The first of the van der Waals parameters is the maximum strain λ_{max} defined by the maximum value of the first strain invariant $I_{1,max}$

$$I_{1,max} = \lambda_{max}^2+2/\lambda_{max}=y=M_c/M_u : \; \lambda_{max}-\sqrt{y} \tag{20}$$

λ_{max} is depends on the chain length parameter y that is defined as number of stretching invariant units of the molecular weight M_u. The second van der Waals parameter a is introduced for describing global interaction between chains.

From equation (16) we deduce the mechanical equation of state expressing the nominal force in the strain mode α as

$$f_e = (\partial W_e / \partial L)_{T,p} = G_o D_e \left[1/(1-\eta_e) - a\sqrt{\phi_e} \right] \tag{21}$$

For our purposes it is sufficient to consider three different modes of strain.

uniaxial:

$$I_{1,u} = \lambda^2 + 2/\lambda ; \ D_u = \lambda - \lambda^{-2} \tag{22}$$

equibiaxial:

$$I_{1,b} = 2\lambda^2 + \lambda^{-4} ; \ D_b = \lambda - \lambda^{-5} \tag{23}$$

simple shear:

$$I_\gamma = 3 + \gamma^2 ; \ D_\gamma = \gamma \tag{24}$$

It is a significant matter that the van der Waals-network parameters M_u, λ_{max} and a are invariant showing no dependence on the type of strain. This is exactly one of the most important reasons for making the van der Waals network in its global regime a conformational gas.

THERMO-ELASTICITY IN VAN DER WAALS NETWORKS
Together with the equation (5) and equation (21) we arrive at

$$(\delta Q / \partial L)_{p,T} = -T(\partial K_e / \partial T)_{p,L} = -T\left[(A_o \partial f_e / \partial T)_{p,L} + f_e (\partial A_o / \partial T) \right] \tag{25}$$

with

$$T[\partial f_e / \partial T]_{p,L} = T f_e \left[\partial \ln G_o / \partial T + \left(\overline{D_e} / D_e + \overline{W_e^*} / W_e^* \right)(\partial \ln L_o / \partial T)_{p,L} \right] \tag{26}$$

and

$$\partial G_o / \partial T = 1/T + \partial \ln \rho / \partial T + \partial \ln <r^2> / \partial T + \partial \ln <r_o^2> / \partial T = 1/T - \beta - \mu \quad (27)$$

The temperature dependence of the length of reference $L_o(T)$ characterised by

$$\left(\partial \ln L_o(T) / \partial T \right)_{p,L} = \beta \tag{28}$$

we formulate

$$\partial \ln \rho / \partial T = -3\beta; \quad \partial \ln <r^2> / \partial T = 2\beta \tag{29}$$

μ as the temperature coefficient of non-iso-energetic rotational isomers is given by

$$\partial \ln <r_o^2> / \partial T = \mu \tag{30}$$

The other derivatives in equation (26) are then written as

$$\overline{D}_u = -\left(\lambda + 2\lambda^{-2}\right); \quad \overline{D}_b = -\left(\lambda + 5\lambda^{-5}\right); \quad \overline{D}_\gamma = -\gamma \tag{31}$$

In addition we have

$$\overline{W}_a^* = \left(\phi_{max}\overline{\phi}_a - \phi_a\overline{\phi_{max}}\right) / \left(2(1-\eta_a)^2\eta_a\phi_{max}^2\right) - a\overline{\phi_a}/2\sqrt{\phi_{max}} \tag{32}$$

with

$$\overline{\phi}_u = -\lambda D_u; \quad \overline{\phi}_b = -2\lambda D_b; \quad \overline{\phi}_\gamma = -\gamma D_\gamma \tag{33}$$

Hence, the formulation of the thermo-elastic relations of a van der Waals network is complete. The equations are based on linear approximations implicating that intrinsic properties even in highly strained networks should uniquely be related to the thermodynamic coefficients in the unstrained system. It is the simplicity of this concept that the knowledge of the coefficient of thermal expansion, the temperature coefficient of the non-iso-energetic rotational isomers and the compressibility should lead to a quantitative understanding.

THE VOLUME EFFECT

According to equation (8) we have to deduce the pressure dependence of the van der Waals-force K_a. Together with equation (21) we are led to

$$\left(\partial V/\partial L\right)_{p,T} = \left(\partial K_a/\partial p\right)_{T,L} = A_o\left(\partial f_a/\partial p\right)_{T,L} + f_a\left(\partial A_o/\partial p\right)_{t,L} \tag{34}$$

with

$$\left(\partial f_a/\partial p\right)_{T,L} = f_a\left[\partial \ln G_o/\partial p + \left(\overline{D_a}/D_a \overline{W_a^*}/W_a^*\right)\left(\partial \ln L_o(T)/\partial p\right)_{T,L}\right] \tag{35}$$

Defining the isothermal compressibility in the unstrained rubber

$$\kappa = -3\left(\partial \ln L_o(p)/\partial p\right)_{T,L} \tag{36}$$

the other derivatives are written as

$$\begin{aligned} \partial \ln \rho/\partial p &= \kappa \\ \partial \ln <r^2>/\partial p &= -2\kappa/3 \end{aligned} \tag{37}$$

Hence, we arrive at

$$\begin{aligned} \left(\partial \ln G_o/\partial p\right)_{T,L} &= \partial \ln \rho/\partial p + \partial \ln <r^2>/\partial p \\ &\quad - \partial \ln <r_o^2>/\partial p \end{aligned} \tag{38}$$

whereby

$$\partial \ln <r_o^2>/\partial p = \mu_p \tag{39}$$

It is an interesting matter that we may find out by describing volume-strain effects whether the equilibrium conformation of chains depends on pressure.

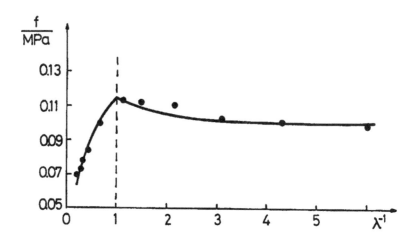

Figure 1., Mooney-plot ($f^*=f/D$ against λ^{-1}) of poly-dimethylsiloxane (PDMS) in simple extension and uniaxial compression [27]. Solid line theoretical [19]

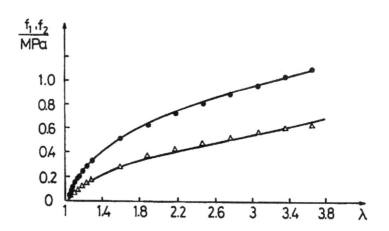

Figure 2., Principal nominal forces f_1 (●) and f_2 (△) of polyisoprene in pure shear ($\lambda_2=1$) [28]. Solid line theoretical (G=0.38 MPa, $\lambda_{max}=18$, a=0.22) [18,19]

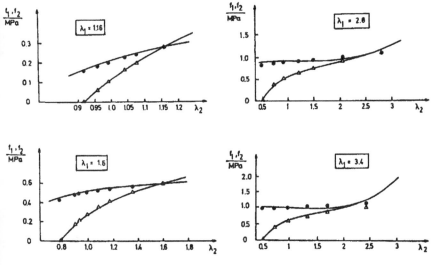

Figure.3, Nominal principal forces f_1 (●) and f_2 (Δ) of polyisoprene in biaxial extension [28]. Solid lines theoretical, parameter as in Fig.(2) [18,19]

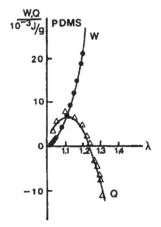

Figure 4., $W(\lambda)$ and $Q(\lambda)$ of PDMS is the mode of simple extension [20]. Lines theoretical ($G=0.36$MPa,$\lambda_{max}=10$,$a=0.2$,$\beta=2.210^{-4}$ K^{-1},$\mu=9.610^{-4}$ K^{-1})

Table 1., Natural Rubber (NR)

M_u=68 g mol^{-1}; λ_{max}=11.2; a=0.28
β=2.4 10^{-4} K^{-1}; μ=9.6 10^{-4} K^{-1}

Table 2., Styrene-Butandiene Rubber (SBR)

M_u=42 g mol^{-1}; λ_{max}=10; a=0.3
β=4 10^{-4} K^{-1}; μ=-4 10^{-4} K^{-1}

Table 3., Natural Rubber (NR): ΔV/V-Measurements

M_u=68 g mol^{-1}; λ_{max}=11.2; a=0.28
κ=8 10^{-4} MPa^{-1}; μ_p=5.3 10^{-4} MPa^{-1}

THE STATE OF REFERENCE

It is possible to describe deformation for different types of strain with the same set of van der Waals network parameters [18,19,16,17]. Representative examples are shown in figure 1 [18], figure 2 [19] and figure 3. The strain-energy is in any case increased by deforming the network, the entropy is diminished (W=-TdS). If the network is stable we find the absolute minimum of strain-energy for all types of strain at $\Sigma\lambda_i^2=1$ (entropy-maximum). In respect to the maximum strain-energy at $\Sigma\lambda_{max,i}^2=y$, the unstrained state is in the centre of symmetry.

THERMO-ELASTICITY

The thermo-elastic inversion of PDMS networks in the range of smaller strains [20] is fairly well fitted by the calculation (figure 4). The description is accurate up to largest strains (figure 5). Natural rubber displays the same behaviour and is equally well understood (figure 6). A negative μ seems to be typical for styrene-butadiene rubber (SBR). In this case the internal energy does not change very much (figure 7). It is possible to describe the classical experiments of Anthony, Caston and Guth [21] (figure 8, Table 4)

Table 4., Parameters for calculating the Anthony et. al. data [30]

M_u=68 g mol^{-1}; λ_{max}=9.3; a=0.27; β=2.6 10^{-4} K^{-1}; μ=6.5 10^{-4} K^{-1}

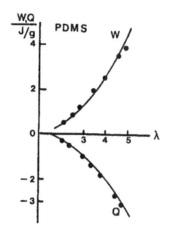

Figure 5., PDMS: W(λ), Q(λ) [2,16]. Solid lines theoretical (G=0.36 MPa; λ_{max}=10, a=0.2, ß=2.2 10^{-4} K^{-1}, μ=9.6 10^{-4} k^{-1}

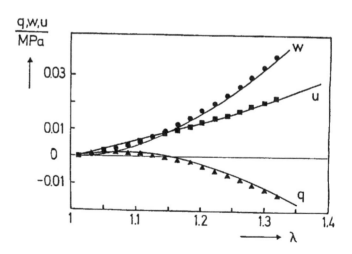

Figure 6., Heat density q(λ), strain-energy density w(λ) and internal energy density u(λ] of NR against λ at 295 K [24]. Solid lines theoretical (parameters Table 1)

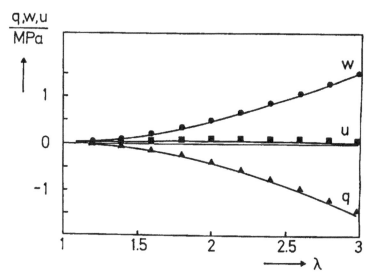

Figure 7., Heat density q, strain-energy density w and internal energy density u of SBR against λ at 295 K [20]. Solid lines theoretical (parameters Table 2)

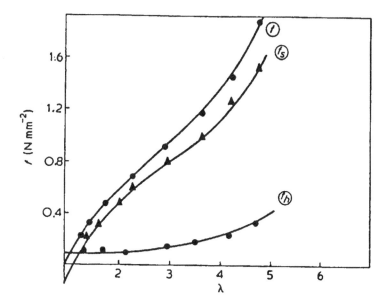

Figure 8.,The force of retraction f and its components as obtained from force-temperature curves at fixed lengths [30]. Solid lines calculated (parameters Table 4)

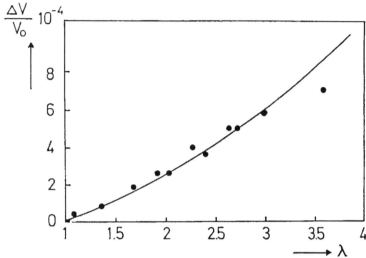

Figure 9., ΔV/V(λ) of NR according to Göritz [21]. Solid line is theoretical, parameters Table 3.

VOLUME EFFECTS

The good quality of the fit of calculations to measurements of $\Delta V(\lambda)/V$ of NR according to Göritz (17] is to be seen by evidence from the plot in figure 9. It is remarkable that the calculation is successful under the condition of

$$\mu_p = -2\kappa/3 \equiv \partial \ln <r^2>/\partial p \tag{40}$$

The memory term of these networks seems not depend on pressure. This is equivalent as to suggest identical pressure-induced changes of conformation of network chains and of the unperturbated chains of reference.

CONCLUSIONS

It is clear that equation (4) is the fundamental equation of a network in the system of coordinates {T,p,L}. The van der Waals-network model provides then a full understanding of the mechanical behaviour in different modes of deformation.

That the thermo-elastic phenomena are also understood proves that internal equilibrium is established. The volume is increased with strain, holes and in-harmonic oscillations are produced as demanded by thermodynamics. The non-iso-energetic rotational isomers are redistributed in unique dependence on the strain whereby this transformation is strictly interrelated with the caloric behaviour.

A real network behaves in the global regime like a weakly interacting van der Waals conformational gas characterized by the structure parameters (M_u, λ_{max}, a). Cooperative processes control the liquid-like inter- and intramolec-ular properties up the largest strains whereby all the necessary changes can be related to the representative thermodynamic coefficients in the unstrained system (β,μ,κ,μ_p). Hence, it demonstrated that is not possible to enforce top-ological changes by stretching a network. Networks remain in the local reg-ime "normal liquids" even when being stretched to very high degrees.

PREDICTIONS

We are encouraged to make some predictions. We have to be aware that the states of reference of the global and the local level are different. While the liqidlike and intramolecular properties are heuristically related to T=0 K and p= 0 MPa, we have to take the unstrained network with the maximum entropy as reference. The symmetry in both levels, the global and the local one, is therefore different. We have the chance to test consequences because of knowing the fundamental equation of the van der Waals-network. We may therefore compute thermo-elastic phenomena and volume changes for each type of strain. From the plot of the relative volume changes as de-picted in figure 10 one learns that the volume is decreased in modes where we have enforced deformation by pressure. On the other hand, the volume increases in extensional modes of deformation. Internal properties display

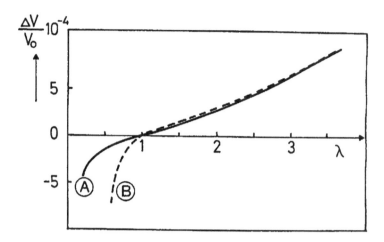

Figure 10., ΔV(λ)/V for (A) uniaxial and (B) equi-biaxial compression and extension computed with the same parameters as depicted in Fig.(8) [24]

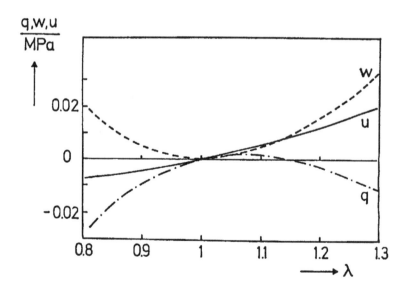

Figure 11: Work- w(λ) and heat-exchange q(λ) and internal energy density u(λ) fore uniaxial extension and compression (parameters Table 4)

analogous anti-symmetries. In pressure-induced modes of deformation there should for example not any more happen any thermo-elastic inversion. In this mode, the internal energy is predicted to decrease with strain while it increases in extensional modes. This illustrated with the plot in figure 11.

One of these anti-symmetries was first reported by Göritz [22] who found that in uniaxial compression a thermoelastic inversion was not any more to be seen.

It is true that the fundamental equation of networks as formulated in equation (4) is correct. Adjustments of internal equilibrium display anti-symmetric effects manifesting the very peculiar way in which the short-range regime is controlled by the global level.

REFERENCES

1) H.B.Callen, Thermodynamics, John Wiley & Sons, New York, London, Sydney, Toppan Comp. Tokyo (1960)

2) H.G.Kilian, Polymer 22, 209 (1981)

3) H.G.Kilian, Makromol. Chem., Macromol. Symp. 40, 185 (1990)

4) G.Winkler, G.Reinecker, M.Schreiber, Europhys. Letters 8, 493 (1983)

5) H.G.Kilian, T.Vilgis, Makromol. Chem. 185, 193 (1984)

6) L.R.G.Treloar, The Physics of Rubber Elasticity, Oxford, At The Clarendon Press (1958)

7) J.E.Mark, B.Erman, Rubberlike Elasticity A Molecular Primer, John Wiley, New York, Chichester, Brisbane, Toronto, Singapore (1988)

8) P.J.Flory, A.Ciferi, C.A.J.Hoeve, J. Polymer. Sci. 45, 235 (1960)

9) P.J.Flory, Trans. Farad. Soc. 57, 829 (1961)

10) G.Allen, M.J.Kirkham, J.Padjet, C.Price, Trans. Farad. Soc. 67, 1268 (1971)

(11)   A.J.Chompff, S.Newman, Polymer Networks, Plenum Press, New York-L
         don (1971)

(12)   H.G.Kilian, Coll. Polym. Sci. 260, 895 (1982)

(13)   H.G.Kilian, T.Vilgis, Makromol. Chem. 185, 193 (1984)

(14)   A.E.Green, J.E.Adkins, Large Elastic Deformations, Clarendon Press (19

(15)   J.E.Mark, Macromol. Rev. 11, 135 (1976)

(16)   M.Zrinyi, H.G.Kilian, F.Horkay, Coll. Polym. Sci. 267,311 (1989)

(17)   M.Zrinyi, H.G.Kilian, F.Horkay, Makromol. Chem., Macromol. Symp. 30:1
         1989)

(18)   H.Pak, P.J.Flory, J. Polym. Sci., Phys. Ed., 17, 1845 (1979)

(19)   S.Kawabata, M.Matsuda, K.Tei, H.Kawai, Macromolecules 14, 154 (1981

(20)   Y.K.Godovsky, Vysokomol. Soed. A19, 2359 (1977)

(21)   R.L.Anthony, R.H.Caston, E.Guth, J. Phys. Chem. 46, 826 (1942)

(22)   D.Göritz, 10th network-group meeting and IUPAC on polymer networks,
         Jerusalem (1990)

Polymer Networks '91 pp. 119-145
Dosek and Kuchanov (Eds)
© VSP 1992

Neutron scattering investigation of the deformation at molecular scales in polymer networks

J. Bastide[§], F. Boué[¶], E. Mendes[§], F. Zielinski[¶], M. Buzier[¶], G. Beinert[§], R.Oeser[*] and C. Lartigue[*]

[§]*Institut Charles Sadron (CNRS; CRM-EAHP), 6 rue Boussingault F-67083 Strasbourg Cedex, France*
[¶]*Laboratoire Léon Brillouin (CEA-CNRS), CEN Saclay F-91191 Gif sur Yvette Cedex, France*
[*]*Institut Laue-Langevin, B.P. 156X, F-38042 Grenoble Cedex, France*

1. INTRODUCTION

The neutron scattering technique is now currently used in order to study polymer networks ("dense" rubbers, or swollen rubbers i.e. gels). The first series of experiments were essentially devoted to the probe, at molecular scales, of the mechanism of network deformation: it was an attempt to check the more directly possible the validity of the hypotheses underlying the rubber elasticity theories [1] (for a review of the main theories see [2-5]). For instance, the swelling of a gel was originally described only on the basis of an *extension* of the elementary strands (or meshes) of the network in the three directions of space [2]. The neutron scattering results modified this vision. Roughly speaking, they led to compare the swelling - at least in the case of a gel prepared in the presence of solvent - to a kind of *unfolding* of the network from a "crumpled" or "interspersed" state at high concentrations of polymer [6,7]. Along this scheme, an eventual extension of the meshes is a second order perturbation resulting from a sort of feed-back of the expansion.

Regarding the uniaxial extension of dense rubbers, the neutron scattering experiments allowed to observe an average orientation of labelled chains chemically bond to the structure, as expected in stretched systems. By selecting the intensities scattered in the direction parallel and perpendicular to the axis of elongation (i.e. for the scattering vectors q oriented respectively parallel and perpendicular to the stretching direction), it was possible to achieve a comparison between calculated form factors and experimental ones. Quite often, however, one found inconsistencies when comparing the data in these principal directions [8-9]. For instance, for very long chains passing through many crosslinks, they fitted approximately some of the calculated curves, but for values of the crosslinking density (introduced as an adjustable parameter) differing with the direction of observation. Instead of restraining the analysis to the directions parallel and perpendicular to the stretching axis, one can choose also other ways for visualizing the data. For instance, when a bidimensional detector is employed (a kind of chess plate with e.g. 64x64 squares, each of them counting the neutrons), one can draw iso-intensity curves on a map of the detector. In the case of elongated rubbers containing labelled chains (connected or not to the network), one finds then surprising anomalies with respect to the expectations. Contrarily to the case of the analysis in definite directions, such anomalies affect then the general shape of the curves. Their existence is thus obvious even on qualitative grounds and this is the reason why we will base the present paper on them.

2. ANOMALIES IN THE ISO-INTENSITY CURVES, FOR ELONGATED RUBBERS

2.1. The "lozenge" trend [8-11,18]

As mentioned above, one of the ways commonly employed for investigating the deformation in polymer networks consists of focusing on systems containing *labelled paths*, i.e. long deuterated chains connected to the network by many junctions. When the sample is strained, the neutron scattering intensity provides an insight into the deformation mechanism at distance scales corresponding to the elementary mesh but also at larger distance scales. In the case of dense rubbers submitted to an imposed uniaxial elongation and studied as a function of the time elapsed after the deformation, a surprising change of shape of the iso-intensity curves has been observed (especially when the

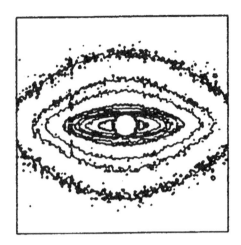

Figure 1: Example of the "lozenge" trend: iso-intensiy curves (different levels) observed in the case of a polystyrene network containing "labelled paths" (deuterated chains). The stretching ratio is equal to 4 and the sample has been elongated at 150°C, the duration T_R of the relaxation in between the stretching and the quenching of the sample at room temperature is approximately equal to 20 s (see below). The stretching axis is vertical. q range: $6 \cdot 10^{-3} < q < 6 \cdot 10^{-2}$ Å^{-1}.

synthesis of the network: The sample used for this experiment was synthesized in the presence of CCl_4 at a polymer volume fraction ϕ_c equal to 0.085 (swelling degree in the reaction bath equal to 11.75). The crosslinking was achieved through the reaction (of Friedel and Crafts type) of CH_2Cl lateral groups, which had been fixed at random on the chains in a previous step. The polymer employed is a mixture of 70% protonated chains ($M_w = 525000$) with 30% of deuterated ones ($M_w = 545000$). Tin tetratchloride was used as a catalyst. After the reaction, the samples were carefully washed and the solvent was extracted. The maximum swelling degree in toluene was found to be equal to 24. For details of the chemistry and more experimental results see refs. 12-13.

stress relaxation procedure: We recall that T_g is approximately equal to 100°C for polystyrene. The samples are stretched at a temperature $T > T_g$ in a silicon oil bath; the elongation is maintained at this temperature during a certain time T_R, called the relaxation time in this context, before quenching the sample by taking it out of the oil bath. The series of experiments are ordinarily performed by varying not only the duration of the relaxation but also the temperature at which it is achieved. The data can be compared using the time-temperature superposition principle [14].

extension ratio λ is large enough; λ>3). In the first part of the relaxation process, the iso-intensity curves are approximately elliptical [8-10], as it is expected from calculations based on the current elasticity theories (when looking in more detail, the way the aspect ratio of the ellipses changes with q does not always fit very precisely the theoretical expectations). At larger relaxation times, however, the iso-intensity lines progressively distort and tend to become diamond-shaped: one calls them sometimes *lozenge patterns* . See an example in Figure 1. A remarkable feature of this distortion is that it occurs essentially at relaxation times larger than the Rouse time of a free chain (in a melt) of size comparable to that of the average elementary chain of the network. This means that the lozenge anomaly appears in a range of times for which the mechanical equilibrium should be reached, according to the classical picture.

Although calculated curves closer to the experimental lozenges have been obtained by several independent methods (refinements or modifications of the current models), the origin of this anomaly is not yet clear. In the following, we will approach this problem from the experimental side by showing that it may have some connection with the even more spectacular anomaly of the butterfly patterns, presented in the next paragraph.

2.2. The "butterfly" patterns [11,15-18]

This anomaly was discovered when studying Polydimethylsiloxane (PDMS) rubbers containing free deuterated PDMS chains, i.e. not chemically linked to the network [15]. Originally, such kind of experiments were undertaken in order to measure which constraint the matrix imposes to the free chains. When the sample is stretched, one expects a significant orientation of the chains at small relaxation times which is effectively observed: one gets elliptical iso-intensity lines with the long axis oriented perpendicular to the stretching direction. At large relaxation times, provided the free chains are not too long with respect to the size of the average elementary chain of the network, one reaches a state of quasi-isotropy: the iso-intensity lines turn back to circles, as before deformation. But, surprisingly, this stage is not the final one. The signal continues to evolve: at low q, the scattering intensity strongly increases in the direction *parallel* to the stretching axis.

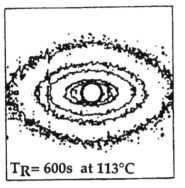

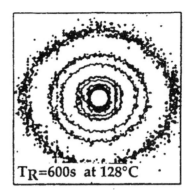

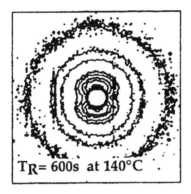

Figure 2: Appearance of butterfly patterns in the iso-intensity lines during the relaxation of an elongated polymer network containing free deuterated chains (M_w=138 000). The stretching ratio λ is the same for the 3 series of curves: λ= 2.14. T_R is the duration of the relaxation in between the stretching (at 113°C) and the quenching at room temperature. The stretching axis is vertical. q range: $6 \cdot 10^{-3} < q < 6 \cdot 10^{-2}$ Å^{-1}.

Fig. 2a: individual orientation of the deuterated probes; Fig. 2b: quasi-isotropy, i.e. relaxation of the individual orientations; Fig. 2c: appearance of the butterfly patterns, visible in the center of the detector.

sample: The network studied here was prepared from a mixture of 10% deuterated polystyrene chains with 90% undeuterated polystyrene chains carrying aminomethyl side groups. These chains were dissolved in toluene at a polymer volume fraction ϕ_c equal to 0.085. The crosslinking was achieved by reaction of the aminomethyl groups with terephtaldialdehyde added to the solution. In order to perform the neutron experiments, the gel was deswollen and then dried in a vacuum oven. In good solvent, the maximum swelling degree Q_{max} these samples was found equal to 16.6. For more data and more experimental details, please see references 12 and 13.

Correspondingly, the iso-intensity lines adopt a double-winged shape with the long axis oriented *parallel* to the elongation direction, which is extremely unusual. See examples in Figures 2 and 3. We have shown elsewhere that the appearance of such *butterfly patterns* originates most presumably in an anisotropic change of the distribution of the labelled free chains in space and not in an individual change of conformation of these chains [16].

We think that the understanding of this spectacular phenomenon is important for a better description of the properties of polymer networks. In the following, we will make a proposal regarding its physical origin with the help of a model developed for the case of swollen gels. But before, we would like to show that the lozenge distortion seems to be related to the question of "butterflies".

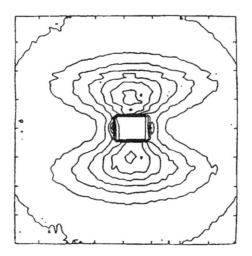

Figure 3: Another example of butterfly patterns studied at smaller q values ($3 \ 10^{-3} < q < 3 \ 10^{-1} \ Å^{-1}$) [12,13]. The sample is of same type as for the experiment reported in Figure 2, except for the crosslinking density, which is larger (the maximum swelling degree Q_{max} in good solvent is approximately equal to 8; the crosslinking reaction presumably proceeded during the drying process). The molecular weight of the labelled chains is the same as above. The stretching ratio was equal to 3 . The sample contained a small amount of solvent in order to decrease the glass transition temperature (which was approximately 50°C); it was stretched at 80°C and T_R was equal to 60s.

2.3. Connection between "butterflies" and "lozenges"

Take now a rubber containing free deuterated chains much longer than the average elementary chain of the network, and stretch it. At low T_R, the iso-intensity lines are again ellipses. But, when T_R is increased at constant elongation, one does not reach a state of quasi-isotropy, as in the previous case (when the labelled chains were shorter). Instead of that, one observes the apparition of "lozenges". As the relaxation proceeds, the anisotropy of these "lozenges" decreases whereas, at the same time, "butterfly patterns" are growing at smaller q values. See Figure 4. In the present situation, the deuterated chains are some kinds of "labelled paths" entangled with many elementary meshes, but not fixed to the network. In the case when the labelled chains are chemically linked to it, the "lozenges" remain permanently (i.e. even for very large relaxation times) [8,11]. Therefore, it may be inferred that ellipses are distorted into lozenges by the physical phenomenon that leads to butterflies when the labelled chains are free, and which cannot occur completely when they are linked to the network.

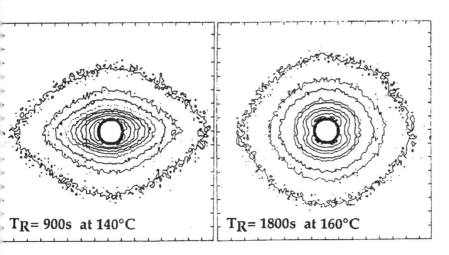

T_R= 900s at 140°C T_R= 1800s at 160°C

Figure 4: Example of a "lozenge" to "butterfly" transition [12,13]. The network has been prepared according to the method summarized in the caption of Fig. 2. The molecular weight of the free deuterated chains is approximately equal to 7.5 10^5. The maximum swelling degree Q_{max} in good solvent is equal to 13, thus the molecular weight of the average mesh of the network is very likely below 35000. In this experiment the stretching ratio λ was equal to 3.

3. HOW BUTTERFLY PATTERNS CAN BE PRODUCED BY "UNSCREENING" OF HIDDEN CLUSTERS

In this section, we are going to deal with ideas which are mostly contained in a model published elsewhere [19,20]. We will not present the quantitative predictions, which can be found in the original papers, but focus on the physical ideas and use a slightly different route to introduce them. We will present also the results of some experimental tests.

Let us assume that a semi-dilute solution of extremely large polymer chains is lightly crosslinked almost instantaneously in a random fashion (with however a large average number of crosslinks per chain such a way that the gel point is widely passed and the sol fraction can be considered as negligible). This means that the links which are established between the chains are arranged in space as the molecules in a perfect gas. As a result some regions are formed which are more dense in crosslinks than the gel on an average (with naturally other regions less dense than the average). Thus, the gel contains in a sense a perfect gas of crosslinking heterogeneities. To go further, we have first to deal with something that may look strange, namely the "structure" of a perfect gas.

3.1. What is the structure of a perfect gas?

For simplicity, we will restrict ourselves to a "lattice gas", where the molecules are placed on the nodes of a regular lattice (each site of this lattice having approximately the same volume as a molecule); this makes more easy the definition of neighbouring molecules but does not modify the arrangement of the matter in space at scales of distance much larger than the size a of the site. At every moment, the structure of this perfect lattice gas can be described as the result of a site percolation process: the occupancy of a site is drawn at random with a probability equal to the ratio p of the number of molecules over the total number of sites [21,22]. It is well known that in a site percolation process, clusters of fist neighbours on the lattice are formed. These clusters are not compact; they are branched, their distribution in sizes is very wide and the larger are self similar with a fractal dimension equal to 5/2 (if the number of occupied sites is enough to reach the vicinity of the percolation threshold). See Figure 5.

lattice gas → no small angle scattering

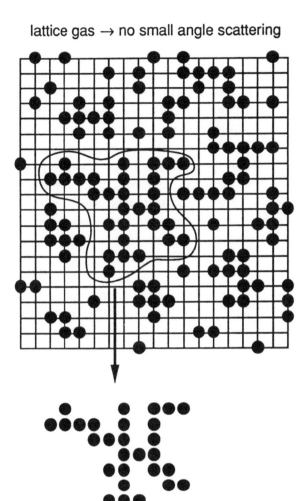

isolated cluster → small angle scattering

Figure 5: Schematic representation of a lattice gas in two dimensions. When the proportion of occupied sites is large enough, clusters of first neighbours are formed. The larger ones are fractal and are partly interpenetrated with smaller ones (percolation model). The larger clusters taken separately should scatter significantly at low q, but the small angle scattering of the ensemble is expected to be flat (in the case of a sufficiently large system) because of the randomness in the arrangement of the elementary molecules (the dots).

The average size of the clusters increases with p and, for p larger than a certain threshold p_c, one of them (at least) becomes "infinite", which means that it reaches the edges of the lattice. Since the points have been put at random on the lattice, the small angle scattering (in practice for q<<1/a; q being the scattering vector amplitude) by this system will be flat: there is no structure factor in a perfect gas. Nevertheless, for p comparable to p_c, this system contains at every moment large objects which, if they were considered separately (i.e. without the isolated points and the small clusters), would lead to a considerable scattering with a strong dependence in q. This apparent paradox is easily resolved by remarking that the correlations between the scattering centres of one given cluster (the intra-cluster correlations) are exactly balanced in a perfect lattice gas by the correlations between the points of this cluster and those of all the other clusters (the inter-cluster correlations).

A comparable type of phenomenon occurs in a semi-dilute solution of long chains: such a system scatters less than a more dilute solution of the same chains (at $c \neq c^*$ for instance, c^* the overlap concentration of the chains). This arises also because the correlations along the same chain in the semi-dilute state are partly masked (screened out) by the correlations with the other chains, as a result of the interpenetration of the coils [4]. In the case of the perfect lattice gas close to the percolation threshold of the occupied site, the intra-cluster correlations are screened out because the bigger clusters contain smaller clusters which themselves contain even smaller cluster etc, in analogy with the fitting in of Russian dolls, as explained by Daoud and Leibler [23].

3.2. An "imaginary experiment" with a particular glass

In the situation of the lattice gas, the screening of the intra-cluster correlations is inevitable. The clusters form and unform continuously; it is obviously impossible to separate and thus to detect them, for example with the help of a scattering method. Let us however imagine an "imaginary experiment" on a very close system, an ideal binary liquid, in order to bring the clusters to light. Assume that two types of small molecules (say "black" and "white" ones) are mixed together at a temperature much larger than the demixion temperature at such a concentration that, at every moment, the molecules of one type (say the black ones) are nearly percolating. In other words, the black molecules are expected to be located at random in this ideal mixture and for this reason form clusters of first neighbours of percolation type. Suppose that an isotopic "contrast" is established between the black molecules and the white ones. Though the contrast, the clusters of black molecules will not be detected in a neutron scattering experiment, because of the overall randomness of the system. Assume now that this liquid is quenched very rapidly, in order to form a glass, such a way that the arrangement in space of the molecules at a given instant is frozen. The clusters of black points will keep the same structure and thus will remain undetectable.

But now, imagine that the glass is stretched, owing to some kind of plastic deformation of the matrix of white molecules. And last but not least, suppose that, due to black-black interactions stronger than the white-white ones, the clusters of black molecules deform less than the white matrix. This means that under strain the inter-cluster distances will be changed more strongly than the intra-cluster ones. As a result the exact balance between the intra-cluster and inter-cluster correlations, characteristic of the thorough randomness, will be destroyed: this should lead to huge changes of the scattering intensity.

Since they have been assumed to deform less than the average sample, the black clusters should separate (i.e. disentangle) in the direction parallel to the stretching, when the sample is elongated. Accordingly, one expects an increase of the scattering intensity for q vectors aligned with the stretching axis (and for low amplitudes of these vectors), since the intra-cluster correlations must be revealed. In the direction perpendicular to the elongation, the sample contracts. Thus, the black clusters should interpenetrate more in this direction. The randomness of their arrangement should also be destroyed along this direction and the scattering intensity is again expected to be enhanced. However, it should increase far less than in the parallel direction, because the deformation ratio is necessarily much lower. For q vectors oriented along an axis of angle θ, intermediate between the parallel and the perpendicular ones, one expects a scattering intensity variation I(q) also intermediate between those corresponding to the principal directions. See Figure 6a. The curves representing I(q) for different angles θ will therefore reach a given level of intensity for q values less high as the orientation draws nearer to the perpendicular axis. A representation of this variation in polar coordinates leads to an iso-intensity line. Note finally that when the reference level I_{ref} is larger than the extrapolated value of the scattering intensity at zero q in the perpendicular direction, the complete iso-intensity curve (over 360°) splits in two lobes: then it is seen that this anisotropic unscreening process leads to the appearance of what we are looking for, namely a "butterfly pattern" . See Figure 6b.

Several types of systems might exhibit behaviours of this kind. One can think first about some metallic or semi-metallic alloys, generally studied for their spin-glass properties. Under stretching, the distribution in space of the clusters of the ferro-magnetic species could be strongly modified and this could change not only the scattering properties (e.g leading to the appearance of butterfly patterns) but also the magnetic ones.

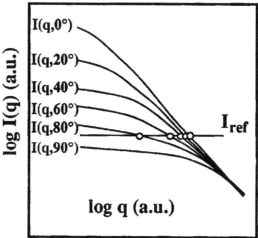

Figure 6a: Schematic representation (log-log plot) of the scattered intensities in directions of various orientations with respect to the stretching axis (including the parallel (0°) and the perpendicular (90° ones). As a result of the "unscreening" of intra-cluster correlations along the direction of elongation, one expects an enhancement of the intensity at low q values, more especially large as the vector q is aligned with the stretching axis (see text). I_{ref} is the reference level which has been chosen for the construction of an iso-intensity line. The reference level i reached for decreasing values of q, as the angle θ with respect to the stretching axis increases. In this example, the reference level is no reached for any q value in the perpendicular direction.

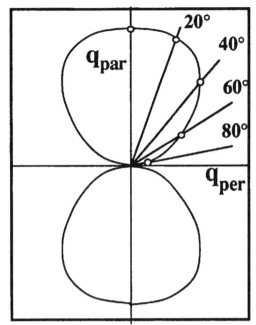

Figure 6b: Schematic representation of the construction, from the curves of Fig. 6a, of an iso-intensty line with a level equal to I_{ref}. In the case represented here, there must be a certain critical angle θ_c such as $I(O, \theta_c) = I_{ref}$. Thus, for $\pi-\theta_c > \theta > \theta_c$, the level I_{ref} is not reached and there is a "forbidden wedge" (of angle $\pi-2\theta$) for the iso-intensity line, which crosses itself at the origin and forms two lobes.

Another candidate could be a rubbery matrix containing mineral charges. In that case, the elementary grain of the charges (consisting of carbon black or silica, for example) would play the role of the black molecules in the above example. It is well known that such grains form in general branched aggregates, that could be partly interpenetrated (even if their structure is not exactly that of percolation clusters). Under stretching of this reinforced sample, one expects - this is again classical - a deformation of the clusters less important than the average one (and, conversely, a deformation of the interstitial matrix larger than the average one). The clusters should then disentangle partly in the direction parallel to the stretching axis and interpenetrate more in the perpendicular direction. As a result, one expects here also the observation of butterfly patterns in a scattering experiment (if a contrast is established between the charges and the matrix). In addition, note that, because of the elongation, the percolation threshold can be passed in the direction perpendicular to the axis of elongation, if, before stretching, the concentration of charges was slightly smaller than that corresponding to the transition.

Various kinds of inhomogeneous polymer networks might also behave in a comparable manner. We can think first about bimodal networks prepared, for example, by end-linking of long and short precursor chains (as done currently by J. Mark et al [24]). If the crosslinking of the long and short molecules is approximately performed at random (in the limits of the constraints imposed by the incompressibility, see the related discussion of the "correlation hole" problem in [4]), and if the proportion of the short ones is large enough, clusters of each species should be formed. Presumably, the clusters of short chains will deform less than the clusters of large ones, as the network is stretched. Thus, if if one type of chains is labelled with respect to the other one, butterfly patterns should be observed, because of the separation process introduced above. We postpone a more detailed discussion of this interesting question of bimodal networks. Another example of intrinsically inhomogeneous networks is provided by the case of statistically crosslinked gels, which initiated this long digression. We are going to discuss this problem in the following. Then, we will try to generalize the arguments to the case of dense rubbers.

3.3. Unscreening of clusters in statistical gels

At the beginning of this section, we have described the crosslinking heterogeneities of statistical gels in their state of preparation as a kind of "perfect gas" of "harder" regions (i.e. of regions which deform less than the gel on an average). According to the foregoing, it is equivalent to say that the structure of the heterogeneities is that of percolation clusters. Consider now the distribution in space of the polymer. Since the presence of the crosslinks induces in general a certain contraction, the polymer concentration in the clusters may be a little larger than the average one (over the whole gel). Because of this "contrast" in polymer concentration, a cluster taken alone should be detected by a neutron scattering experiment (if the polymer is marked with respect to the solvent). However, as above in the case of the lattice gas, intra-cluster correlations are expected here to be totally screened out by inter-cluster correlations.

Thus, the cluster structure should remain, in the state of preparation, undetected by a scattering experiment. Note that for the same reasons, the polymer density correlation length ξ should remain nearly identical to that of the semi-dilute solution of the same concentration (the preexisting one).

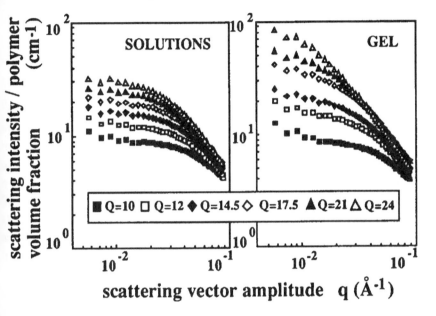

Figure 7: Scattering intensity divided by the polymer volume fraction ϕ, for a statistical gel at different degrees of swelling Q and for a series of semi-dilute solutions of polymer fractions equal to those of the gel (log-log plot) [27]. It can be observed that the signal of the gel is approximately the same as the one of the solution, at the concentration of preparation of the network (Q=10). On the other hand, as the swelling degree Q is increased, the signal of the gel over-tops more and more, at low q, the one of the equivalent solution. The contrast is achieved by swelling the network or dissolving the chains in deuterated toluene.
sample: The gel studied here was prepared using a Friedel-Crafts reaction at a concentration $\phi=0.1$ in 1,2-dichloroethane. The crosslinker was 1,4-bis(chloromethyl)benzene, added at a concentration of 0.8% of mole per mole of monomer. The gels were washed, dried and then reswollen in deuterated toluene for the neutron scattering experiment [25-27].

As in the above example of the strained glass (or the reinforced rubber), it is easy to predict that this exact balance is extremely unstable. It should be destroyed by any type of deformation of the gel. A change of the swelling degree or simply of the shape of such a network should result in the cancellation of the "absolute" randomness in the arrangement of the crosslinking heterogeneities. The regions rich in junctions should indeed expand less than the gel on an average, if solvent is added to the network, and also should be stretched less than the average if the system is elongated. One expects then important changes in the scattering intensity upon deformation of the network, as a result of the unscreening of the intra-cluster correlations.

Under swelling, since the more crosslinked clusters are supposed to absorb less solvent than the interstitial medium (the less crosslinked regions), they should separate from each other and disentangle isotropically. At the same time, the "contrast" of polymer concentration between the clusters and the rest of the gel should be enhanced. Because of these two effects, one expects a strong increase of the scattering intensity at small scattering vector amplitudes (small q) when the swelling degree is increased (i.e. when the polymer concentration is decreased). The correlation length in the gel is also expected to raise strongly with the swelling degree and should become, in the overswollen network, much larger than in semi-dilute solution of the same concentration. Such behaviours have effectively been observed in the case of gels prepared by random crosslinking (using a Friedel and Craft crosslinking reaction) of a semi-dilute solution of large molecular weight polystyrene chains. An example concerning the scattering intensity is shown in Figure 7; we refer to [25-27] for experimental details and for a more complete presentation of the results (including the variation of the correlation length).

In the case of an elongated gel, the situation should be very close to that described above in the example of the strained glass. Since they are expected to deform less than the average sample, the clusters should separate (i.e. disentangle) in the direction parallel to the stretching, when the gel is elongated. They should become partly "visible" and the scattering intensity for q vectors aligned with the stretching axis should thus increase strongly for low amplitudes of these vectors. The correlation length ξ is expected to be changed into an anisotropic quantity and should be raised significantly in this direction. Since its volume is conserved, the sample is contracted in the direction perpendicular to the elongation. Thus, the clusters should interpenetrate more in this direction.

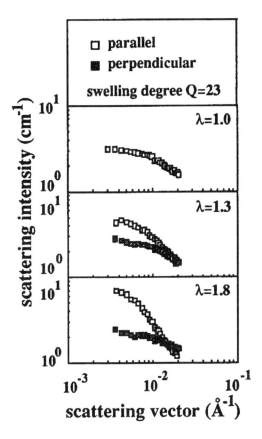

Figure 8: Scattering intensity (log-log plot) in the directions parallel and perpendicular to the elongation axis, for a statistical gel (analogous to that of Fig. 7) stretched by different factors λ [27]. An isotropic stray signal has been subtracted at very low q. The network was approximately in its state of maximum swelling in deuterated toluene (Q=23).

If the sample is stretched in its prepared state, the scattering should also increase because the destruction of the randomness of the cluster structure, but far less than in the parallel direction. For the same reasons as before, one should then observe some butterfly patterns for relatively low scattering vector amplitudes. If the gel is elongated at a swelling degree larger than that corresponding to the state of preparation, the contraction of the sample in the perpendicular direction may be compared to a deswelling back to the state of preparation. As a result one should then observe a decrease of the scattering intensity in this direction and the butterfly patterns should be even more pronounced.

Statistical gels nearly identical to those employed for the swelling experiments reported above have been submitted to a uniaxial elongation, at swelling degrees larger than that of perparation. All the features expected from the model have been observed: *butterfly patterns,* strong increase of both the scattering intensity at low q and the correlation length in the direction parallel to the stretching, decrease of both the scattering intensity at low q and the correlation length in the perpendicular direction [25-27]. Some examples of these behaviours are shown in Figs. 8 and 9. Please refer to [25-27] for a more complete presentation of the data.

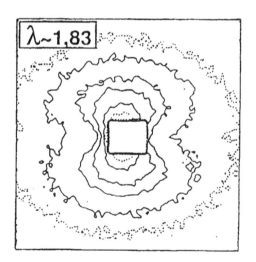

Figure 9. Iso-intensity curves (different levels) obtained in the case of a statistical gel stretched in the vertical direction by a factor $\lambda=1.83$ (same series of experiments as in Fig. 8; Q=23) [27]. q range: $3 \ 10^{-3} < q < 3 \ 10^{-1} \ Å^{-1}$.

In summary, according to the model presented above, some heterogeneities having the shape of interpenetrated clusters can be rather important in size and in consequences and remain, though, nearly undetectable because they are formed totally at random. However, as soon as the randomness is destroyed, for example by a change of shape of the sample, the existence of these heterogeneities is easily detected. In the case of the statistical gels submitted to a uniaxial elongation, a line of argument of this type leads to an anisotropic unmasking of the clusters. It predicts the appearance of spectacular *butterfly patterns* in the neutron scattering iso-intensity curves. Such patterns have effectively been observed.

4. A PROPOSAL FOR THE EXPLANATION OF BUTTERFLY PATTERNS IN DENSE NETWORKS

When comparing Figure 9 with Figure 3, it is clear that the butterfly patterns observed in the case of highly swollen gels and those obtained for small chains trapped in dense rubbers are surprisingly alike. Then it becomes a question to know whether or not the physical origin of the phenomenon is the same in both cases. There is not yet a definite answer to this question. Several theories have already been proposed that might explain - directly or indirectly - the butterfly phenomenon in dense systems [28-33] and some experimental tests are still underway. Thus, the proposal of related origins that we are going to make now must, for the moment, be considered as a guess.

The starting point is not really an assumption, but more an experimental fact: most of the *real* polymer networks are, whatever the synthesis method, inhomogeneous to some extent [34-39,7]. The reasons for the formation of heterogeneities are numerous, but we think that some of the features described in the case of random crosslinking of semi-dilute solutions can be found in many types of systems, including dense rubbers. Due to the randomness always present in the gelation processes, one can indeed expect the formation of regions more crosslinked than the average, having the shape of non-compact i.e. branched clusters. The size distribution of these "harder" regions in expected to be very large. In the general case, the characteristics of such clusters should not be exactly those of percolation animals; however they might be, in many cases, partly interpenetrated.

4.1. Non-uniform swelling of a rubber by a polymer melt

Assume now that such a piece of rubber is put in contact with a
melt of polymer chains of the same chemical nature as the network
chains (with eventually an isotopic difference like deuteration in order to
get a contrast for neutron scattering experiments; in most cases the
enthalpy of mixing associated with the isotopic differences is low enough
to be negligible). For simplicity, we postulate that the polymerization
index N_f of these chains is comparable with that of the average mesh (N_c)
of the rubbery network (say $0.3\ N_c < N_f < 3\ N_c$). These chains will behave
as a polymeric solvent and penetrate, to some extent, into the rubber. It is
however important to realize that the maximum swelling degree of a
network immersed in such a melt is generally much smaller than in a
good solvent of low molecular weight [40]. It varies very strongly with
both the molecular weight of the chains and the crosslinking density of
the network. This is due to the fact that the term in favour of the
expansion of the network, the entropy of mixing of the polymer chains, is

Figure 10: Schematic representation of an inhomogeneous dense
network, slightly swollen with free deuterated chains. In this figure, we
have assumed, for simplicity, that the free chains do not penetrate in the
more crosslinked clusters.

Figure 10a: Nearly random clusters coexisting with a nearly random
arrangement of free chains. The clusters are almost "invisible" for the
scattering.

inversely proportional to their molecular weight and thus generally very weak. Therefore, if a certain quantity of a melt is absorbed by the rubber, it should not be evenly distributed in space. The local concentration of free chains should, on the contrary, vary strongly with the local crosslinking density, the "harder" clusters containing far less mobile polymer than the looser ones. Such a situation is represented schematically in Figure 10a.

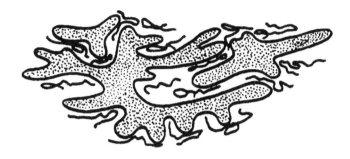

Figure 10b: Small relaxation times T_R: the small chains are oriented and the positions of their centres of gravity are transformed affinely with the macroscopic length scales. The clusters are also deformed affinely up to rather short lenght scales. A conventional anisotropy is detected in a scattering experiment.

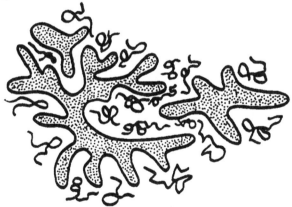

Figure 10c: Large relaxation times T_R. Relaxation of the orientation of the free chains and partial contraction of the clusters.The positions of the free chains are rearranged, according to the new shape of the less crosslinked regions, in which they concentrate. The clusters are labelled by the free chains and their structure is now partly (and anisotropically) unscreened because the randomness has been destroyed. Butterfly patterns should be observed at low q values.

4.2. Idealized case of a rubber crosslinked in the bulk, with random heterogeneities

We suppose that "harder" clusters having a totally random structure have been formed, as the result of the crosslinking process of a rubber in the bulk. Besides, we assume that a small amount of free deuterated chains has been spread on a piece of this network and have penetrated in it (the average concentration of free chains could be, say, 10 or 20%). We consider now the neutron scattering properties of such a sample. Theoretically, the observed signal can be expressed as a convolution of the "form factor" of the labelled chains by an inter-chain term (a sort of "structure factor") characteristic of their distribution in space. Here, the distribution is not uniform because of the presence of the heterogeneities. Thus, the inter-chain contribution should, in principle, have a detectable effect on the scattering; a way to figure out what this structure factor consists of is to remark that the harder clusters structure must be revealed through the "contrast" created by the difference of local density in free chains. In other words, the structure of the clusters and their arrangement in space is expected to influence the shape of the scattering envelope (essentially in the range of scattering vector amplitude such as $qR_g<1$, R_g being the radius of gyration of the labelled chains). Note however that the contribution of this structure factor should remain weak in the present case, for the same reasons as in the foregoing example of a gel in its prepared state: the clusters remained undetectable because they were formed at random. Here, the swelling degree of the network resulting from the penetration of the free chains is small, since the content of free chains is postulated to be low. The clusters should therefore be only slightly separated (and very weakly deformed) with respect to the state of thorough randomness, in which they are "invisible". Therefore, one expects only a slight enhancement of the scattering intensity, at low q values, with respect to the case of a random dispersion of the same labelled chains in a uniform melt.

4.3. Unmasking of the clusters resulting from the deformation and the relaxation of the rubber

Let us deform uniaxially our heterogeneous sample containing free labelled chains. Immediately after the elongation has been settled, the distances between pairs of scattering centres should be changed affinely in the macroscopic strain (except for very close points along the same chain). See Figure 10b. As a result, one expects to probe, at very small relaxation times, essentially the effects of the individual orientation of the labelled chains: this is the behaviour classically observed in the case of stretched samples, with, particularly, elliptical iso-intensity curves. After some time has elapsed at constant macroscopic strain, a reorganization of the local deformation should occur. The first "natural" evolution is the return to isotropy of the free chains and a certain loss of affineness of the network chains (the current models of rubber elasticity do not agree entirely about the extent of this last process). This should lead to iso-intensity lines becoming nearly circular curves again, after a relaxation time of the order of that of the Rouse time of the free chains. Because of the fluctuations in space of the density of junctions, one expects then an additional relaxation consisting of a contraction of the more crosslinked clusters, to the prejudice of the less crosslinked ones, which should, on the contrary, be strained more. This process should be accompanied by a change of the arrangement in space of the free chains. They should still be preferably located in the "softer" regions and, accordingly, the new shape of the clusters structure is expected to be marked. See Figure 10c. Note that this re-distribution of the free chains concentration is governed by their diffusion constant, and thus strongly dependent on their molecular weight.

In this regime of relaxation times, there should not be any noticeable change of the "form factor" of the mobile chains: they should stay isotropic. On the other hand, one expects a strong modification of their "structure factor". In order to estimate it qualitatively, we can reason again upside down, on the basis of the structure of the "harder" clusters which hinder a uniform distribution of the labelled molecules. Since the clusters have been assumed to be deformed less, after relaxation, than the sample on an average, they can be expected to separate (disentangle) noticeably in the direction parallel to the elongation axis. Thus, in this direction, the effect of the stretching should be comparable to that of a swelling: it should partly "unscreen" the correlations associated with the cluster structure and the scattering intensity should increase. On the opposite, the scattering intensity should decrease in the direction perpendicular to the stretching since the clusters which were slightly separated by the penetration of the free chains should re-interpenetrate more. Such a scenario is, in its principle, the same as the one we have encountered for the elongation of swollen gels. Therefore, one expects, as in this previous case, the appearance of iso-intensity lines having the shape of butterfly patterns. According to this line of arguments, the resemblance of the curves in these *a priori* different situations would not be fortuitous and the physics behind the phenomena would be very similar.

4.4. "Real" systems

Real rubbers are not necessarily crosslinked in bulk, they can be prepared in the presence of solvent and subsequently deswollen. Moreover, the labelled mobile chains may be present at the time of the crosslinking and not added afterwards (as in the experiments presented in Section II). We do not think that these modifications should, in general, invalidate the reasoning given above. The important elements are the cluster structure of the heterogeneities and the fact that they are partly interpenetrated. A deswelling preceding the stretching should essentially enhance the interpenetration of the clusters, if such regions of higher crosslinking density are formed in course of the gelation process. The distribution of the free chains should also be perturbed by the presence of these clusters, even if these mobile molecules were present before the gelation. The only noticeable difference that we see would be a non-random structure of the clusters before stretching. Thus the scattering by the free chains in such rubbers should be neatly different from that by the same chains in a melt. But the principle of an anisotropic further unscreening of the cluster structure should be conserved and the butterfly patterns should again be observed.

5. SUMMARY AND CONCLUDING REMARKS

In summary, this paper has been devoted to the possible role of crosslinking heterogeneities in the deviations, with respect to theoretical expectations, which are currently found when analysing data of neutron scattering by polymer networks. We have focused our interest on the phenomenon of "butterfly patterns" appearing in the iso-intensity lines in the case of stretched networks containing uncrosslinked labelled probes (solvent or chains). We recall how, in the case of the entirely labelled gels (i.e. networks swollen with a deuterated solvent), the butterfly patterns seem to be the consequence of the presence of heterogeneities having the structure of partly interpenetrated clusters. Then, we propose a generalization of such a mechanism for the case of dense rubbers containing uncrosslinked labelled chains. Crosslinking heterogeneities having the structure of partly interpenetrated clusters might also exist in such systems and they might be "labelled" by the free chains behaving as a kind of polymer solvent. This would be enough to explain the observation of butterfly patterns in the case of a uniaxial deformation. The "lozenge" trend observed in the case of labelled chains fixed to the network by many crosslinks might result from fluctuations of the local deformation oɪ these labelled paths, because of the presence of the heterogeneities.

The line of arguments that has been presented is based on an idea which is contained in the percolation model: random processes often form clusters. Although some of them can be very big, these clusters are not easily detected. Taken individually, they should lead to a noticeable scattering (if a contrast is established between them and the surrounding medium). But the randomness of their formation makes them totally "invisible", for the scattering, when they are studied as a whole. This surprising disappearance results from an exact cancellation of the intra-cluster correlations by the inter-cluster correlations, which is obvious when remembering that the clusters have been formed by points located at random. Such a subtle balance must be destroyed as soon as a deformation, acting differently on the clusters than on the rest of the sample, is imposed to the system. According to this picture, the butterfly patterns result from nothing more than an anisotropic unmasking of the cluster structure. A mechanism of comparable type may happen in various systems, and not only in the case of inhomogeneous networks. In this context, one may wonder whether the butterfly patterns recently observed by T. Hashimoto and K. Fujioka in sheared semi-dilute solutions can be understood with such an idea of cluster separation[41]. Even more generally, it would not be surprising if some unexplained strongly non-linear phenomena originated from a revelation of clusters formed by a random process.

ACKNOWLEDGEMENTS: We are indebted to L. Leibler, T. Hashimoto, P. Higgs, T. Vilgis and H. Benoît for fruitful discussions We would like also to thank M. Antonietti for having suggested the use of terephtaldialdehyde as crosslinker. The neutron scattering experiments have been performed at Institut Laue-Langevin and at Laboratoire Léon Brillouin; we would like to thank P. Lindner and A. Brulet for their valuable help..

REFERENCES:

[1] see for example:
  H. Benoît, D. Decker, R. Duplessix, C. Picot, P. Rempp, J. P. Cotton, B. Farnoux, J. Polymer Science, 14, 2119 (1976).
  J. A. Hinkley, C. C. Han, B. Mozer and H. Yu, Macromolecules, 11, 836 (1978)
  S. B. Clough, A. Maconnachie, G. Allen, Macromolecules, 13, 774-775 (1980)
  M. Beltzung, C. Picot, P. Rempp, J. Herz, Macromolecules, 15, 1594 (1982).
  E. Geissler, A. M. Hecht, R. Duplessix, Macromolecules,16 ,712 (1983)
  J. Bastide, R. Duplessix, C. Picot, S. Candau, Macromolecules, 17, 83 (1984)
[2] P. J. Flory, Principles of Polymer Chemistry, Cornell University Press, Ithaca, N. Y. (1953)
[3] R. T. Deam, S. F. Edwards, Phil. Trans. Royal Society of London, Ser. A, 280, 317 (1976).
[4] P. G. de Gennes, Scaling Concepts in Polymer Physics, Cornell University Press, Ithaca, N. Y. (1979)
[5] B. Erman, P. J. Flory, Macromolecules, 15, 800 (1982).
  B. Erman, P. J. Flory, Macromolecules, 15, 806 (1982).
[6] J. Bastide, S. J. Candau, C. Picot, J. Macromol. Phys., B19, 13 (1981)
[7] S. J. Candau, J. Bastide, M. Delsanti, Adv. Polym. Sci., 44, 27 (1982).
[8] J. Bastide, F. Boué, Physica A, 140, 251 (1986)
[9] F. Boué, J. Bastide, M. Buzier, C. Collette, A. Lapp, J. Herz, Prog. Coll.Polym. Sci, 75, 152 (1987)
[10] F. Boué, Adv. Polym. Sci., 82, 47 (1988)
[11] F. Boué, J. Bastide, M. Buzier, in Molecular Basis of Polymer Networks, A. Baungärtner, C. Picot, Eds., Springer Verlag, Berlin (1989)
[12] F. Zielinski, Thesis, Université Paris VI (1991)
[13] F. Zielinski et al. to be published
[14] J. D. Ferry, Viscoelastic properties of Polymers, Wiley, New York (1980)
[15] R. Oeser, C. Picot, J. Herz, in Polymer Motion in Dense Systems,

D. Richter, T. Springer, Eds., Springer Verlag, Berlin (1988).

[16] J. Bastide, M. Buzier, F. Boué, in Polymer Motion in Dense Systems, D. Richter, T. Springer, Eds., Springer Verlag, Berlim, 112 (1988).

[17] J. L. Barea, R. Muller, C. Picot, in Polymer Motion in Dense Systems, D. Richter, T. Springer, Eds., Springer Verlag, Berlin, 93 (1988)

[18] F. Boué, J. Bastide, M. Buzier, A. Lapp, J. Herz, T. A. Vilgis, Coll. Polym. Sci., 269, 195-216 (1991)

[19] J. Bastide, L. Leibler, Macromolecules, 21, 2647 (1988).

[20] J. Bastide, L. Leibler, J. Prost, Macromolecules, 23, 1821 (1990)

[21] D. Stauffer, Introduction to Percolation Theory, Taylor and Francis, London and Philadelphia (1985)

[22] D. Stauffer, A. Coniglio and M. Adam, Adv. Polym. Sci., 44, 105 (1982)

[23] M. Daoud, L. Leibler, Macromolecules, 21, 1497 (1988)

[24] J. E. Mark, Adv. Polym. Sci., 44, 1 (1982)

[25] E. Mendes, P. Lindner, M. Buzier, F. Boué, J. Bastide, Phys. Rev. Lett, 66, 1595-8 (1991)

[26] J. Bastide, E. Mendes, F. Boué, M. Buzier, P. Lindner, Makromol. Chem., Makromol. Symp. 40, 81-99 (1990)

[27] E. Mendes, Thesis, Université Louis Pasteur, Strasbourg (1991)

[28] F. Brochard-Wyart, P. G. de Gennes, C. R. Acad. Sci. Ser. 2, 306, 699 (1988)

[29] N. Pistoor, K. Binder, Coll. and Polym. Sci., 266, 132 (1988).

[30] S. F. Edwards, T.C.B. Mc Leish, J. Chem. Phys.,92, 6855-7 (1990)

[31] M. G. Brereton, T. A. Vilgis, F. Boué, Macromolecules, 22, 4051 (1989)

[32] A. Onuki, in Dynamics of Ordering Process in Condensed Matter, S.Komura and H.Furukawa (Eds.), Plenum Press, N.Y. (1988).
A. Onuki, in Space Time Organisation in Macromolecular Fluids, F. Tanaka, T. Ohta and M. Doi (Eds.), Springer Verlag Berlin, 94 (1989) and more recent works to be published.

[33] P. Higgs , to be published

[34] N. Weiss, A. Silberberg,, Br. Polym. J., 9, 144 (1977); N. Weiss, T. Van Vliet, A. Silberberg, J. Polym. Sci. - Polym. Phys. Ed., 17, 2229 (1979); A. Silberberg, in Biological and Synthetic Networks; O. Kramer, (Ed.), Elsevier Applied Science, Amsterdam, (1988).

[35] R. S. Stein, J. Polymer Science, Part B, 7, 657 (1969).

[36] K. L. Wun, W. Prins, J. Polymer Science, 12, 533 (1974)

[37] E. Geissler, A. M. Hecht, R. Duplessix, J. Polym. Sci., 44, 30 (1982)
A. M. Hecht, R. Duplessix, E. Geissler, Macromolecules, 18, 2167- 2173 (1985)

[38] J. Bastide, F. Boué, M. Buzier, in Molecular Basis of Polymer Networks, A. Baumgärtner and C. Picot, (Eds.), Springer Verlag, Berlin, 48 (1989).

[39] S.J.Candau, C. Y. Young, T. Tanaka, P. Lemaréchal, J. Bastide, J. Chem. Phys., 17, 83 (1979)

[40] F. Brochard, J. Physique (France), 42, 505 (1981).

[41]T. Hashimoto and K.Fujioka, J. Phys. Soc. Jap., 60, 356-359 (1991)

Polymer Networks '91 pp. 147-157
Dosek and Kuchanov (Eds)
© VSP 1992

Orientation-stress relation of polymer fluids, networks and liquid crystals, subjected to uniaxial deformation

A. Ziabicki

Polish Academy of Sciences, Institute of Fundamental Technological Research, 21 Swietokrzyska Street, Warsaw, Poland

ABSTRACT

"Stress-optical law" which stipulates proportionality between the stress tensor and optical anisotropy in polymer solids and fluids is not a universal principle. Unique relation between stress and optical tensor is justified only for systems, for which stress, and birefringence are controlled by the same, configurational mechanism. Linear behaviour is confined to the range of small stress. Therefore, linear "stress-optical" relations are commonly observed in configuration-controlled solutions, melts and networks of flexible polymer chains, but fail for elasticity-, and dissipation-controlled solutions and suspensions of rigid particles. Comparison of birefringence of various materials in various deformation (flow) regimes reflects as much material properties, as deformation conditions.

A special, simple kind of deformation - uniaxial extension (or steady extensional flow) - reduces relations between stress and birefringence (orientation) tensors to scalar functions. Such characteristics are usually unique, but non-linear, and play important role in the theory of formation highly ordered structures.

STRESS, ORIENTATION, AND OPTICAL PROPERTIES

In various materials subjected to deformation and flow one can observe optical birefringence proportional to stress [1]. Rubber networks, inorganic glasses and crystals subjected to small deformations, polymer solutions and melts composed of flexible macromolecules flowing at not too high deformation rates, provide typical examples. On the other hand, some other materials, such as suspensions and liquid crystals, do not show unique relations between stress and optical anisotropy.

"Stress-optical law" derived from empirical observations consists in linear relation between the traceless tensor of birefringence, **N**, characterizing optical properties, and deviator of the stress tensor **p**. The proportionality constant is often called "stress-optical coefficient"

$$(1) \qquad N = C_{opt} (p - \frac{1}{3} trp\ I)$$

A more general formulation admits N as a non-linear function of the stress tensor, **p**

$$(2) \qquad N = F(p) - \frac{1}{3} trF\ I$$

which, after expansion and application of the Cayley-Hamilton principle reduces to

$$(3) \qquad N = f_1(I_1,I_2,I_3)(p - \frac{1}{3} trp\ I) + f_2(I_1,I_2,I_3)(p^2 - \frac{1}{3} trp^2 I)$$

I_1, I_2, I_3 are three independent invariants of the stress tensor, and I is unit tensor.

In the following, we will discuss stress-orientation problems in somewhat different terms. Instead of optical tensor, **N**, we will analyse orientation tensor, **A**, a characteristic of molecular configuration, directly related to deformation and flow. Components of this tensor yield orientation factors (order parameters) proportional to second Legendre polynomials. On the other hand, the optical tensor, **N**, can be determined by various physical effects and assume different values at the same state of molecular configuration. One of the contributions (intrinsic birefringence) is directly proportional to **A**, but others, like effect of form, make optical tensor much more complex.

Various theoretical analyses of the stress-orientation problem [2-5] indicate that linear (or even unique) relations between stress and orientation can be justified only, when specific assumptions are made. It is evident that "stress-optical law" is by no means universal, and should be applied with great care.

Unique orientation-stress relations can be expected in those systems, in which both stress and orientation are controlled by one mechanism. Such a mechanism can be provided by molecular configura-

tion and elastic response to molecular deformation. Equilibrium
stress in ideal rubbers, and stress in flowing polymer solutions and
melts, results from deformation of flexible macromolecules. In the
case of crystals and glasses, the same role is played by displace-
ment of atoms and molecules under external deformation. Whenever
there appear elastic (configurational) and viscous (rate) mechanisms, stress becomes a function of deformation and deformation
rate, while orientation, as a rule, remains a configuration characteristic. Consequently, stress and orientation are decoupled, and no
unique relation between **A** and **p** is justified. This eliminates from
the materials which obey "stress-optical law" suspensions and solutions of rigid molecules. We have demonstrated [4] that in suspensions of rigid ellipsoidal particles, various components of the
stress tensor produce different orientation effects.

The fact that "stress-optical law" is not universal, does not
exclude its more restricted applications. An example of such application is comparison of orientation-stress characteristics for simple geometry of deformation. Comparison of the behaviour in uniaxial
extension reveals intrinsic material properties, and may contribute
to rationalization of polymer processing.

ORIENTATION-STRESS BEHAVIOUR IN UNIAXIAL DEFORMATION

In fluids subjected to steady extensional flow, uniaxial symmetry reduces deformation rate tensor (velocity gradient) to the
simple form

$$(4) \qquad \dot{e} = \begin{bmatrix} -\frac{1}{2}q & 0 & 0 \\ 0 & -\frac{1}{2}q & 0 \\ 0 & 0 & q \end{bmatrix}$$

identical with that of the deviatoric stress

(5) $$p - \frac{1}{3} \text{trp } I = \begin{bmatrix} -\frac{1}{2}\Delta p & 0 & 0 \\ 0 & -\frac{1}{2}\Delta p & \\ 0 & 0 & \Delta p \end{bmatrix}$$

and the tensor of orientation

(6) $$A = \begin{bmatrix} -\frac{1}{2}\Delta A & 0 & 0 \\ 0 & -\frac{1}{2}\Delta A & 0 \\ 0 & 0 & \Delta A \end{bmatrix}$$

What is different, is only scalar parameters: q, Δp, and ΔA. Since steady-state flow is completely characterized by elongation rate, q, both normal stress difference, Δp, and axial orientation factor, ΔA, can be sought as functions of the single variable, q

(7) $\Delta p = f(q)$

(8) $\Delta A = g(q)$

The above relations are not true for unsteady flow of visco-elastic fluids, where time is an additional variable.

The mechanisms responsible for eqs. (7) and (8) need not be identical and may be quite complex. In viscoelastic systems, stress can be controlled by dissipative and elastic mechanisms, while orientation "feels" only elastic (configurational) effects. However, in steady extensional flow, a single variable reflects all effects present in the fluid. Functions $f(q)$ and $g(q)$ are usually monotonical, which leads to a unique relation between Δp and ΔA

(9) $\Delta A = h(\Delta p)$

Realizing that all stress and orientation effects disappear in the absence of flow

$$(10) \qquad \Delta p(q=0) = \Delta A(q=0) = 0$$

we arrive at the expansion

$$(11) \quad \Delta A = (g'/f')_{q=0} \, \Delta p + \frac{1}{2} \, [(g''f' - g'f'')/(f')^3]_{q=0} \, (\Delta p)^2 + \ldots$$

where $f^{(n)} = d^n \Delta p/dq^n$, and $g^{(n)} = d^n \Delta A/dq^n$.

A similar reasoning for uniaxial deformation of solid materials is based on deformation tensor, e, as an origin of stress and orientation

$$(12) \qquad e = \begin{bmatrix} \lambda^{-1/2} & 0 & 0 \\ 0 & \lambda^{-1/2} & 0 \\ 0 & 0 & \lambda \end{bmatrix}$$

with linear elongation, λ, as a scalar characteristic.

Orientation cannot be a linear function in the entire range of stresses. Stress in the deformed, or flowing material can increase infinitely, while orientation is naturally limited by the ideal order. Consequently, a reasonable form of the orientation-stress relation (9) is one, starting as a linear function of Δp at small stresses, and asymptotically levelling off at high Δp, approaching a limit ΔA_{max} corresponding to ideal orientation

$$(13) \qquad \Delta A = C_{or} \, \Delta p \qquad\qquad \text{at } \Delta p \to 0$$

$$(14) \qquad \left. \begin{array}{l} \Delta A = \text{const.} = \Delta A_{max} \\ d\Delta A/d\Delta p = 0 \end{array} \right\} \qquad \text{at } \Delta p \to \infty$$

Comparison of orientation-stress characteristics of different materials in arbitrary deformations (eqs. 1-3), does not make much sense unless we are confined to the class of configuration-controlled systems. The results can be due as much to material properties, as to deformation conditions. On the other hand, simple deformation geometry permits to compare scalar orientation-stress functions (eqs. 10,11) of very different materials, reflecting their intrinsic properties.

ORIENTATION-STRESS RELATIONS AND FORMATION OF HIGH-PERFORMANCE MATERIALS

Production of polymer materials with extremely high modulus and tenacity requires formation of molecular structure with high degree of order. The most important element of this order is molecular orientation consisting in parallel arrangement of molecules (molecular segments) as a result of an appropriate mechanical treatment. Formation of uniaxially oriented fibres, rods and films for reinforcement of composites, involves solid-state, or fluid-state processing in the geometry resembling uniaxial extension (eqs. 4 and 12). It has been noted [6-8] that molecular orientation developed in fluid-state processing (fibre spinning, film casting) is largely controlled by orientation-stress characteristics of the fluid. The steeper is increase of orientation with increasing stress, the easier it is to produce high degree of order required for high mechanical performance. Dramatic difference between orientation-stress behaviour of lyotropic liquid crystals on one hand, and polymer melts composed of flexible chains on the other one, explains why aramide fibres can be effectively oriented in solutions, while processing of flexible-chain polyethylene requires solid-state deformation (drawing) [7,8].

ORIENTATION-STRESS RELATIONS FOR VARIOUS MOLECULAR SYSTEMS

Dilute solutions and melts containing non-entangled, <u>flexible macromolecules</u> yield orientation-stress characteristics shown in figure 1. The common mechanism involved in orientation and stress is deformation of non-Gaussian, freely-jointed chains, and related entropic elasticity. Deformation of the chains is not affine, but constant stress is assumed [9]. In the range of small stresses, orientation increases linearly, but levels off at very high stresses.

$$(15) \qquad \Delta A = (v_0/5kT)[\Delta p + a_2(N)(\Delta p)^2 + a_3(N)(\Delta p)^3 + \dots]$$

v_0 denotes volume of the statistical chain segment, and N – number of such segments in the chain. Note, that the initial slope ("orientation-stress coefficient") is independent of molecular weight (N). Higher expansion coefficients (a_2, a_3, ...) become independent of molecular weight above N = 100, yielding limiting orientation-stress curve, characteristic of a single chain. It is interesting to note,

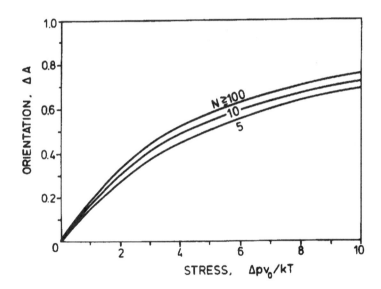

Figure 1. Axial orientation, ΔA, in a system of flexible chains in steady extensional flow vs. reduced normal stress difference, $\Delta p v_0/kT$ [9]. Number of chain segments, N, indicated.

that <u>entanglement</u> of flexible chains, modifying both, stress-deformation and orientation-deformation characteristics, does not change the initial orientation-stress coefficient [10]. An important feature of the orientation behaviour of flexible-chain systems is smooth increase with stress and low sensitivity to molecular weight, limited to very short chains. High degrees of orientation may requi-re application of stress exceeding strength of the material.

Orientation and stress in a network composed of flexible, non-Gaussian chains is similar to that in solutions and melts, and con-trolled by the same mechanism. The presence of permanent con-straints (crosslinks) reduces the average orientation. The results shown in figure 2 were calculated [9] using the model of "modified affine deformation" proposed by Treloar and Riding [11]. The beha-viour at small deformations, and the initial slope are identical with those in uncrosslinked melts and solutions. Different is asym-ptotic behaviour: at high stresses, orientation level reaches only 0.26 - 0.27, rather than 1.00. In unconstrained systems all chains can approach ultimate deformation and orientation; the presence of crosslinks permits full orien-tation of chains parallel to deforma-tion axis forcing others to assume perpendicular orientation which reduces the average degree of segmental orientation, ΔA.

The behaviour of dilute solutions and suspensions of rigid, elongated molecules (rods, ellipsoids), considerably differs from

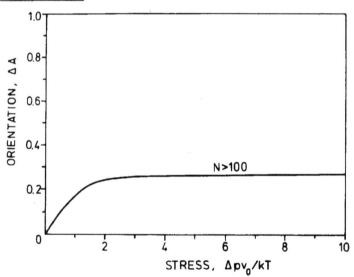

Figure 2. Molecular orientation, ΔA, in a permanent network compos-ed of flexible chains, subjected to uniaxial elongation, vs. reduced normal stress difference, $\Delta p v_0 / kT$. "Modified affine deformation" [9, 11]. N > 100.

that of flexible chains (figure 3). Orientation gradually increases
with stress, and is insensitive to polymer concentration but, unlike
in solutions of flexible chains, the slope of the orientation curve
increases with molecular weight, $N=v/v_0$, and molecular shape (aspect
ratio, $p=L/d$) [4,7].

$$(16) \quad \Delta A = (Nv_0/kT) \ f(p)[\Delta p + b_2(p,N)(\Delta p)^2 + b_3(p,N)(\Delta p)^3 + \ldots \ ]$$

Choosing rigid, large, and elongated molecules, one can produce
high degrees of orientation at reasonably low stresses.

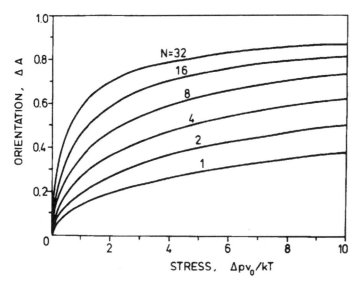

Figure 3. Molecular orientation, ΔA, in a dilute solution of rigid
ellipsoids (p=4) in steady extensional flow vs. reduced normal
stress difference, $\Delta pv_0/kT$ [4]. Molecular size, N, indicated.

In suspensions, the mechanisms involved in stress and orientation
are different. Orientation is a purely configurational property,
while stress results both from configuration and frictional interactions with viscous solvent.

Potential interactions in concentrated solutions or suspensions
of rigid, rod-like particles, introduce an additional mechanism to
stress, and lead to steep, concentration-dependent orientation-
stress characteristics (figure 4). After reaching some critical

concentration (and related intensity of interactions), liquid-
crystalline (nematic) structures are formed [12]. Using Maier-Saupe
[13] interaction potential, $U \cos^2 \vartheta$, proportional to concentration,
one obtains the nematic transition at $U/kT = 7.5$ [14]. In nematic
systems orientation-stress characteristics change dramatically.
Infinitely small stress produces a "jump" in orientation from zero
to a finite value. This initial orientation level, dependent on
concentration and concentration-controlled interactions, results
from interactions alone, and the role of external stress is to determine direction of the orientation axis. Stress improves orientation, which asymptotically approaches unity. Moderately concentrated
solutions of rigid aramides, or p-phenyleno-benzo-bis-thiazole,
which form nematic structures, have been effectively used for spinning highly oriented, stiff and strong fibers.

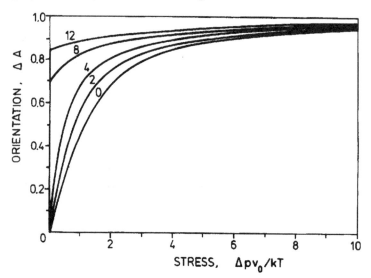

Figure 4. Axial orientation, ΔA, in concentrated solutions of rigid
rods, interacting with a Maier-Saupe potential, in steady extensional flow vs. reduced normal stress difference, $\Delta pv /kT$ [14]. Interaction potential, U/kT, indicated.

REFERENCES

[1] H. Janeschitz-Kriegl, Polymer Melt Rheology and Flow Birefringence, Springer Verlag, Berlin, 1983.

[2] B. D. Coleman, E. E. H. Dill, R. R. A. Toupin, Arch. Rational Mech. Anal., **39**, 358, (1970).

[3] M. Doi, US-Japan Polymer Symposium, October 1985; A. Onuki, M. Doi, J. Chem. Phys. **85** 1190, (1986).

[4] A. Ziabicki, L. Jarecki, Second Conference of European Rheologists, Prague, 1986; Rheol. Acta, **26**, (suppl.), 83, (1988).

[5] A. Ziabicki, in: Third European Rheology Conference (D. R. Olivier, Ed.), Elsevier-Applied Science Publishers, London-New York 1990, pp. 534-536.

[6] A. Ziabicki, J. Soc. Fiber Ind. Japan, **26**, 147, (1970).

[7] A. Ziabicki, in: Polymers for Advanced Structures (M. Lewin, Ed.), VCH Publishers, New York, 1988, pp. 580-601.

[8] A. Ziabicki, Proc. Fiber Producer Conference, Greenville, SC, 1991; Clemson University, Clemson SC, 1991, pp. 1.1-1.8.

[9] A. Ziabicki, L. Jarecki, Colloid and Polymer Sci., **264**, 343, (1986).

[10] M. Kość, Colloid and Polymer Sci., in press (1991).

[11] L. R. G. Treloar, G. Riding, Proc. Roy. Soc. (London), **A369**, 261, (1979).

[12] L. Onsager, Ann. New York Acad. Sci., **51**, 627, (1949).

[13] W. Maier, A. Z. Saupe, Z. Naturforsch., **A14**, 882, (1959).

[14] A. Ziabicki, 22-nd Europhysics Conference on Macromolecular Physics, Structure Formation in Polymer Solutions, Leuven, Belgium, September 1989, A. Ziabicki, to be published.

Polymer Networks '91 pp. 159-166
Dosek and Kuchanov (Eds)
© VSP 1992

Dynamic shear compliance of IR-networks in dependence on crosslink density and filler content

W. Pechhold

Abt. Angewandte Physik, Universität Ulm, Germany

A study of C-C-crosslinked Polyisoprene networks

Samples of Polyisoprene (Cariflex IR 305) were crosslinked by radiation or Dicumylperoxide (DCUP) in a wide range of XL-densities. Shear compliancies in the frequency range 10^{-4} to 100 Hz at different temperatures were measured and summarized into a mastercurve and an activation curve for each sample [1].

The following results are obtained:

a) The mastercurve of both series can be decomposed into two relaxation processes, (i) glass relaxation with its plateau compliance J_{eN} and (ii) shearband relaxation with strength ΔJ_B.

b) JeN is exponentially reduced by effective crosslinks only ($p^*_c = p_c/30$), as understood by the meander model (Fig. 1).

c) The saturation in J_{eN} for higher DCUP-crosslinking (which does not appear with radiation) may be caused by lack of crosslinks across the meander interfaces (preferentially decorated by DCUP-molecules).

d) Using a tight fold length $1/p_f°$-75 monomers for IR (from R_g-data, the total crosslink density p_c (per monomer is determined from ΔJ_B to be $p_c = 2.4/\cdot10^{-2}$. Dose/MGy and pc = $0.97.10^{-2}$ DCUP phr respectively (Fig. 2). This implies that a dose of 0.4 MGy (40 Mrad) is equivalent to 1 part DCUP phr in crosslinking Polyisoprene.

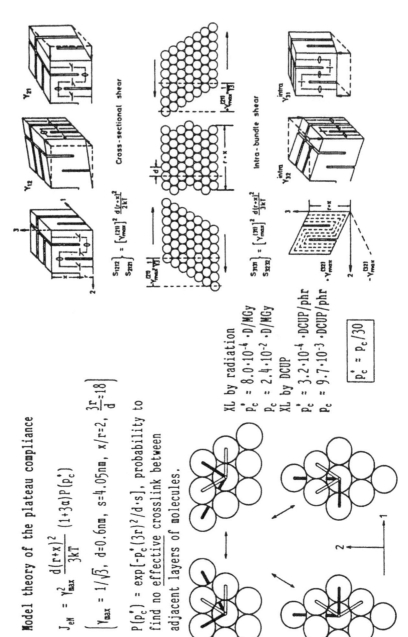

Model theory of the plateau compliance

$$J_{eN} = \gamma_{max}^2 \, \frac{d(r+x)^2}{3kT} \, (1+3\alpha)P(p_c')$$

$$\left[\gamma_{max} = 1/\sqrt{3}, \; d=0.6\,nm, \; s=4.05\,nm, \; x/r=2, \; \frac{3r}{d} \approx 18\right]$$

$P(p_c') = \exp[-p_c'(3r)^2/d \cdot s]$, probability to find no effective crosslink between adjacent layers of molecules.

XL by radiation
$p_c^* = 8.0 \cdot 10^{-4} \cdot D/MGy$
$p_c = 2.4 \cdot 10^{-2} \cdot D/MGy$
XL by DCUP
$p_c^* = 3.2 \cdot 10^{-4} \cdot DCUP/phr$
$p_c = 9.7 \cdot 10^{-3} \cdot DCUP/phr$

$$\boxed{p_c^* = p_c/30}$$

Cross-sectional shear

$$\left.\begin{matrix}S_{1212}\\S_{2121}\end{matrix}\right\} = [\gamma_{max}^{(21)}]^2 \frac{d(r+x)^2}{2kT}$$

Intra-bundle shear

$$\left.\begin{matrix}S_{3131}\\S_{3232}\end{matrix}\right\} = [\gamma_{max}^{(31)}]^2 \frac{d(r+x)^2}{3kT}$$

Fig.1: Discussion of the plateau compliance in the meander model:
Shear deformation modes of the meander cube, mediated by dislocation motion (right side).
Plateau compliance and its exponential reduction by effective crosslinks of density p_c^* (left side).

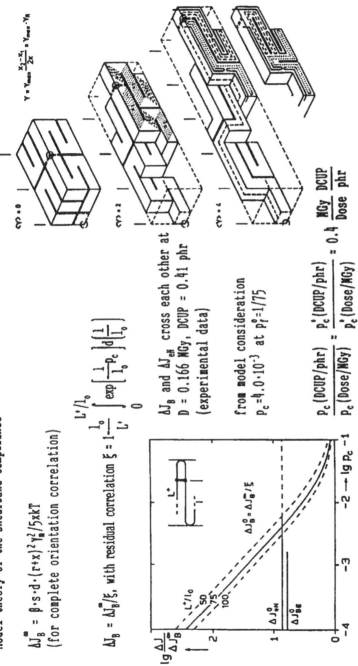

Model theory of the shearband compliance

$$\Delta J_B^\infty = \beta \cdot s \cdot d \cdot (r+x)^2 \gamma_B^2 / 5 \times kT$$
(for complete orientation correlation)

$$\Delta J_B = \Delta J_B^\infty / \xi, \text{ with residual correlation } \xi = 1 - \frac{l_o}{L'} \int_0^{L'/l_o} \exp\left[\frac{1}{I_o} - P_c\right] \left| d\left(\frac{1}{I_o}\right)\right|$$

ΔJ_B and ΔJ_{eN} cross each other at
$D = 0.166$ MGy, DCUP = 0.41 phr
(experimental data)

from model consideration
$p_c = 4.0 \cdot 10^{-3}$ at $p_f^0 = 1/75$

$$\frac{p_c (\text{DCUP/phr})}{p_c (\text{Dose/MGy})} \approx \frac{p_c' (\text{DCUP/phr})}{p_c' (\text{Dose/MGy})} \approx 0.4 \; \frac{\text{MGy}}{\text{Dose}} \frac{\text{DCUP}}{\text{phr}}$$

$$Y = Y_{max} \frac{x_2 - x_1}{2x} = Y_{max} \cdot Y_R$$

Fig.2: Discussion of the shearband compliance in the meander model:
View of a double cube (from a file across a coarse grain) in different states of shearband deformation (right side).
Shearband compliance and its linear reduction by the total crosslink density p_c (left side).

e)   From activation-analysis (Fig. 3) it follows, that its molecular parameters
       ($\Delta \varepsilon^s$, $\Delta Q \gamma$) vary with the square of p_c, as does the change ΔT_g in glass
       temperature (from DSC-measurements). This can be understood in the
       dislocation concept of glass relaxation [2] by assuming, that only double
       crosslinks (statistically generated at the same monomer) increase the energy
       of formation of a segment dislocation.

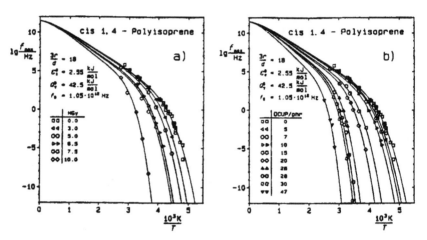

Fig.3a+b: Activation diagrams of two IR-series crosslinked a) by radiation and
          b) by OCUP. The parameters kept constant in each set are specified.

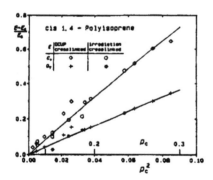

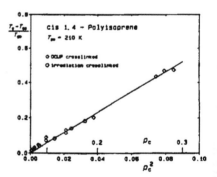

Fig.3c: Relative changes of the free
        energy of formation ($(\varepsilon_s-\varepsilon_{s0})/\varepsilon_{s0}$) of a segment-dislocation
        and of the activation energy
        ($(Q_\gamma-Q_{\gamma 0})/Q_{\gamma 0}$) for jumping of
        segment-dislocations plotted
        vs. the square of the cross-
        link-density p_c^2.

Fig.3d: Relative change of the glass
        temperature ($(T_g-T_{g0})/T_{g0}$) (from
        DSC-measurements) plotted vs.
        p_c^2.

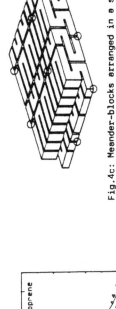

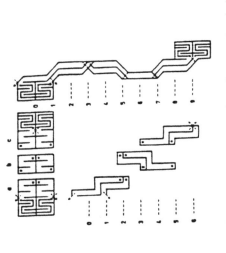

Fig.4c: Meander-blocks arranged in a shearband: every 4 adjacent cubes form the representative element for shearband deformation in melts and rubbers. It is joint to the matrix by 4-bundle-junctions on top and bottom (circles)

Fig.4d: Maximum shear deformation of a meander double-cube by unfolding its superstructure in the 3 bundle-layers

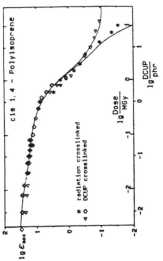

Fig.4a: strain at break in dependence on dose/MGy and DCUP phr

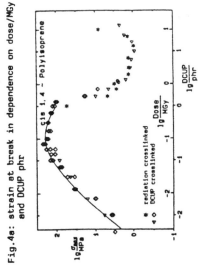

Fig.4b: True stress at break in dependence on dose/MGy and DCUP phr

The ultimate properties, strain and stress at break in tension, were investigated on samples of both series. They also show strong similarity (0.4MGy = 1phr DCUP) except at high crosslinking, where the maximum strain tends to zero for the radiated samples while that of the DCUP-crosslinked samples stay at 10 p.c. (Fig. 4) To interpret quantitatively the ultimate properties in the meander model, it is assumed, that break will occur, if one stretched chain per bundle cross-section exceeds its tensile strength (σ_B = 4.8GPa). The maximum elongation (λ = 35) at low crosslinking is consistent with the unfolding of the meander-superstructure together with that of the tight folded molecules within the bundles.

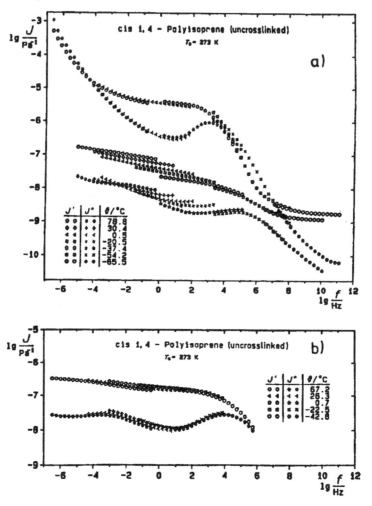

Fig.5: Shear compliance mastercurves of uncrosslinked Polyisoprene:
a) unfilled matrix (upper curves) and filled with φ_F = 0.33 per volume carbon black N220 (lower curves).
b) sample filled with φ_F = 0.21 Vulkasil S.

Dynamic shear compliance of Isoprene Rubber in dependence on filler content and particle size

On 3 series of Polyisoprenes (Cariflex IR 305), filled with different amounts of carbon black (Corax N 220, 330, 660) the dynamic shear compliance was measured in the frequency range 10^{-4} to 100 Hz with small amplitudes at different temperatures. Mastercurves have been composed but should be plotted without the ordinate shifts (Fig. 5), to show its strong differences on filler contents and the type of filler: the negative shift factors ($\sim 1/T$) of the matrix become increasingly positive for carbon black samples ($\phi_F > 0.15$), whereas a white filler (Vulkasil S) does not change the shift factor of the matrix at all. Those dependencies of the shear compliance of filled rubbers on the filler content, particle size and temperature can be understood in a cooperative pair theory (Fig. 6): single filler particles and mobile pairs (having bound

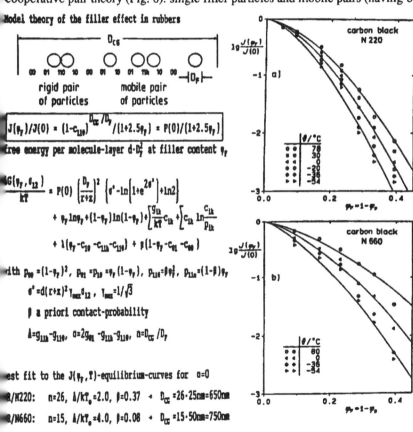

Fig.6: Cooperative pair theory for filler particles in a rubber matrix to describe the shear compliance in dependence on filler content ϕ_F, particle size and on temperature.

rubber in between) do not prevent Intrameander-shear across a coarse grain (D_{CG} ≈700 nm), but rigid pairs (with particle contact) behave as dumbbells or even particle chain and impede rubberelasticity. For carbon black the concentration of mobile pairs increase with temperature, whereas the Vulkasil S particles prefer firm contacts in IR.

The above mentioned theory can also be applied to the stress-amplitude dependence of the shear compliance and to the problem of electrical conductivity of filled rubbers, both areas being still under investigation.

REFERENCES

[1]   Pechhold W., Grassl O., v. Soden W, Colloid Polymer Sci. **268** 1089-1107 (1990)
[2]   Pechhold W, Böhm M, v. Soden, Prog. Colloid Polymer Sci. **75** 23-44 (1987)

Polymer Networks '91 pp. 167-182
Dosek and Kuchanov (Eds)
© VSP 1992

Anomalous properties of hypercrosslinked polystyrene networks

V.A. Davankov and M.P. Tsyurupa

Institute of Organo-Element Compounds, Russian Academy of Science, Moscow 117813, Russia

INTRODUCTION

Crosslinked styrene polymers and, in particular, copolymers of styrene with divinylbenzene (DVB) belong to most thoroughly investigated three-dimensional polymeric networks. Manifold variants of these polymers allow one to easily follow the interrelations between the conditions of the network formation, on one hand, and their structure and properties, on the other.

All known approaches to synthesizing crosslinked polystyrene result in formation of two main types of networks.

The first type is represented by homogeneous (one-phase) networks. Their formation proceeds without any micro-phase separation of the initial homogeneous system, which is a solution. Such polymers are obtained by copolymerization of styrene with DVB in the absence of any diluent [1] or in the presence of a good solvent for polystyrene, such as benzene or toluene, and provided that the amount of DVB does not exceed 10-12 % [2]. One-phase networks can also be obtained by binding in solution reactive ends of polystyrene chains to a polyfunctional node [3] or by partial crosslinking of dissolved polystyrene, i. g., in the course of its polymer-analogous transformation [4]. In the latter case, crosslinking can be accompanied by a certain macro-syneresis of the solvent, but no micro-phase separation takes place [5].

When in dry state, homogeneous polystyrene networks represent non-porous glassy transparent materials with a density close to that of pure polystyrene (1.04 g/cm^3). They do not exhibit any noticeable supramolecular inner structure. These materials swell with thermodynamically good solvents, only, which are able to compensate for the strong chain-to-chain interactions. With the degree of crosslinking rising, the swelling ability drops mono-tonously, whereas the glass transition temperature rises.

The second type of networks, heterogeneous one, represents materials formed under conditions of micro-phase separation of the initial solution. There are three following cases known for such processes:

- Crosslinking copolymerization in the presence of a diluent which dissolves the co-monomers, but precipitates the growing polymeric chains (micro-phase separation according to the mechanism of $\varkappa$-syneresis). In addition to well-known formation of macroporous styrene-DVB copo-lymers [6], crosslinking copolymerization of acrylic acid solutions in acetic acid [7] should be mentioned here.

- Divinylbenzene polymerization (or its copolymerization with styrene) in the presence of large amounts of a good solvent. Here, highly crosslinked poly-DVB micro-gel particles are formed in the system. Having low swelling ability, they do not accumulate the whole amount of the initial solvent (ν-syneresis). After the agglomeration of the micro-gel species takes place, the excess of the solvent remains included in the network in the form of a highly dispersed separate phase [8].

- Crosslinking polymer-analogous transformation of a po-lymer in a medium which is a good solvent for the initial polymer, but a bad one for the final polymer. This is the case with an aqueous polyvinyl alcohol solution reacting with formaldehyde [9].

The micro-phase separation manifests itself in that

the initially transparent solution produces non-transparent gels.

Heterogeneous networks are also formed by styrene-divinylbenzene copolymerization in the presence of dissolved linear polystyrene [10]. Strangely enough, the added polystyrene coils and the growing copolymer coils do not penetrate into each other to a significant extent. The added polymer can be extracted from the final gel almost completely, leaving holes of relatively large diameters.

If the heterogeneous networks are rigid and highly crosslinked, they retain their two-phase nature in the dry state, as well. The permanent pores of these materials can be occupied by any bad solvent. Good solvents, just as is the case with homogeneous networks, cause an additional volume increase of the polymers, which is the less significant, the higher the crosslinking degree and rigidity of the network.

Hypercrosslinked polystyrene, dealt with in the present review-paper, differs principally both from homogeneous and heterogeneous networks.

SYNTHESIS OF HYPERCROSSLINKED POLYSTYRENE

Hypercrosslinked styrene polymers are obtained by crosslinking
(i) linear polystyrene dissolved in dichloroethane or
(ii) slightly crosslinked styrene-DVB copolymers (0.3-2.0 % DVB) swollen with the same solvent.

As the bifunctional or trifunctional crosslinking agents, following compounds can be used: 1,4-bis-(chloromethyl)-diphenyl (CMDP), p-xylilene-dichloride (XDC), tris-(chloromethyl)-mesitylene (CMM), monochlorodimethyl ether (MCDE), dimethylformal (DMF), p,p'-bis-(chloromethyl)-1,4-diphenyl-butane (DPB). In the presence of stannic chloride, they easily react with polystyrene according to Friedel-Crafts reaction and form cross-

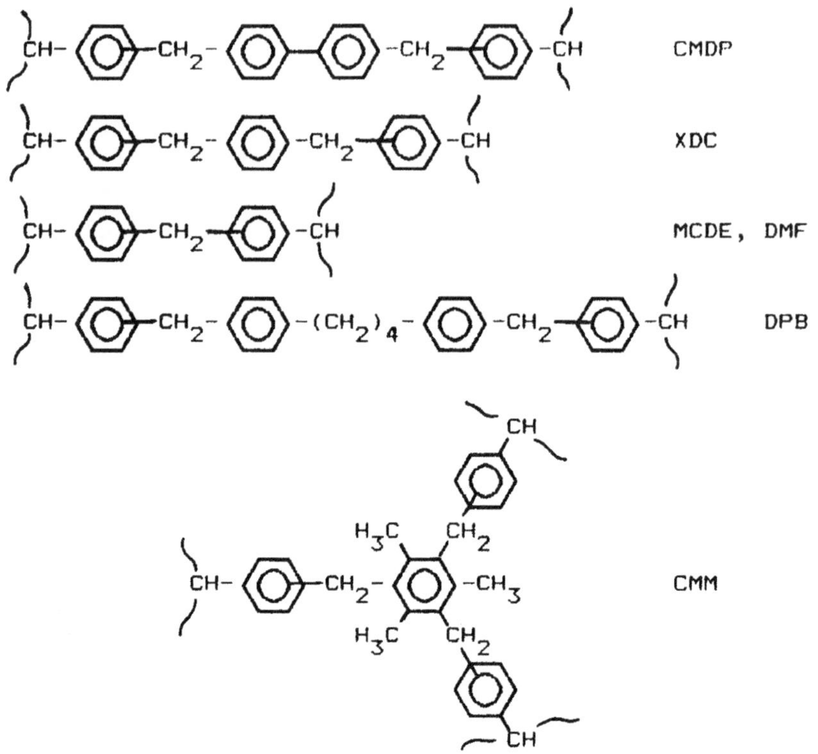

Fig. 1.   Chemical structure of cross bridges

bridges of the structures represented in Fig. 1.

The cross-bridges are relatively rigid, especially
that formed by CMM which simultaneously links three
polystyrene chains. The only   exception   represents   DPB,
since its   four   methylene   groups   are   conformationally
flexible thus allowing easy spatial rearrangements of the
whole network.

Of the above crosslinking agents, CMDP   is   the   most
reactive one. Its complete   conversion   with   polystyrene
requires 4 hours at 60°C in the presence of   0.2   mol   of
stannic chloride per 1 mol of the cross-linker. Reactions
with other compounds require 10 hours, 80°C and 1 mol   of
the catalyst. In all cases, nevertheless,   conversion   of
the initial solution of the polymer and cross-agent   into

a gel starts immediately after adding the catalyst. Therefore, to get sufficient time for preparing homogeneous solution of all the components, it should be cooled to $-20\ ^{o}C$ before adding the catalyst.

The conversion of the polymer solution into a fully transparent and highly swollen with dichloroethane gel is accompanied by partial exclusion of the solvent from the gel phase (macro-syneresis amounts to 20-50 % of the initial volume of the mixture). After the completion of the reaction, the final block of the gel is subjected to disintegration and the irregular particles thus obtained are washed with organic solvents and dried. Spherical particles are obtained by crosslinking beads of styrene-DVB copolymers swollen with a solution of the cross-agent (usually MCDE or DMF) in dichloroethane.

In all systems examined, conversion of the crosslinking agent was observed to be practically complete: according to chromatographic tests, no cross-agents are left in the mother liquid and almost no unreacted chlorine can be found in the final polymer. Therefore, the degree of crosslinking of the network can be formally calculated from the composition of the starting mixture as the ratio of cross-bridges formed to its sum with unreacted styrene units. In the case that 1 ground-mol of polystyrene is reacted with 0.5 mol of a bifunctional agent, the crosslinking degree amounts to 100 %, implying that each of the starting phenyl rings should be involved into formation of cross-bridges.

Networks crosslinked to 40 %, or more densely, should be referred to as hypercrosslinked, since their structure and properties were found to differ drastically from that of known types of polystyrene networks.

PROPERTIES OF HYPERCROSSLINKED NETWORKS

Fig. 2 visualizes the unusual dependence on the

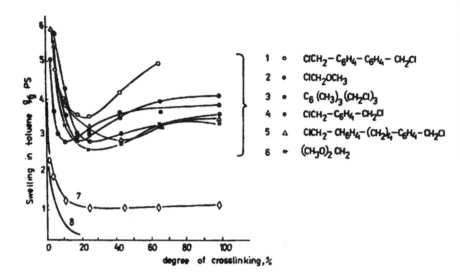

Fig. 2. Swelling capacity in toluene (calculated per
1 g of initial polystyrene) of products crosslinked by
CMDP (1), MCDE (2, 7), CCM (3), XDC (4), DPB (5), DMF
(6), and DVB (8) as a function of crosslinking degree.
Networks based on linear polystyrene (1-6); material
based on beads of a styrene-1%-DVB copolymer (7); conventional styrene-DVB copolymers (8).

crosslinking degree of the swelling ability in toluene
for networks obtained by crosslinking linear polystyrene
with various bifunctional reagents and that of styrene-
DVB(1 %) copolymers crosslinked additionally with MCDE.
As predicted by modern theories of swelling, conventional
styrene-DVB copolymers exhibit a monotonous decrease in
swelling ability with the DVB content rising. On the contrary, the toluene uptake of crosslinked linear polystyrene falls up to a crosslinking degree of 18-25 % only,
but then rises again reaching (at the ultimate crosslinking of 100 %) values of 3-5 ml/g which are characteristic of slightly crosslinked copolymers containing less

than 1 % DVB. When styrene-DVB(1 %) copolymers are cross-
linked additionally in a swollen state, the toluene uptake of the products reaches its lower limit at the bridges
content of 25 % and then remains constant (at a level of
2about 1 ml/g).

It is also evident from Fig. 2 that neither type nor
length of cross-bridges influence the swelling behavior.
Thus, using trifunctional CMM results in networks of
higher swelling capacity as compared to products of
crosslinking with bifunctional reagents. MCDE and DMF
give rise to polymers with equal cross-bridges structure
(of diphenylmethane type), but strongly differing in the
swelling ability.

Surprisingly, correlation was found to exist [11] between the swelling capacity of hypercrosslinked polymers
and reaction activity of the crosslinking agent. The rate
of HCl evolution from the reacting mixtures decreases in
the following series of cross-agents: CMDP >> MCDE > XDC
= DPB. Swelling of the polymers falls in the same sequence. Substituting dichloroethane for nitrobenzene enhances
both the rate of reaction and swelling of polymers produced. From this point of view, it also appears logical
that adding larger amounts of crosslinking reagent and
catalyst (in order to enhance the crosslinking degree)
results in both accelerating the gel formation and enhan-
cing its swelling ability (Fig. 2).

Extremely peculiar appears the ability of hypercrosslinked polystyrene to swell, i. e., increase its volume
in any media including liquids that precipitate polystyrene from its solutions (Table 1). This property is
especially characteristic of densely crosslinked polymers
which swell to equal extents both in good (toluene) and
bad (methanol, hexane) solvents. In the latter case, the
the swelling should be considered as an equilibrium property, as well, since its extent does not depend on the

Table 1, Swelling capacity (ml/g) of products of crosslinking linear polystyrene with xylilene dichloride (XDC), as measured by weighing of dry and swollen material.

| Solvent | Degree of crosslinking, % | | | | |
	11	25	43	66	100
Water	0.05	0.10	0.30	0.80	0.84
Methanol	0.15	0.32	1.13	1.71	1.67
Ethanol	0.10	0.29	1.15	1.86	1.77
Propanol	0.11	0.26	1.43	1.99	1.78
Isopropanol	0.07	0.25	1.23	1.83	1.76
Butanol	0.08	0.22	1.52	2.00	1.79
Isobutanol	0.08	0.11	1.25	1.88	1.72
Heptane	0.16	0.58	1.70	2.00	1.81
Benzene	4,00	2.65	2.23	2.21	1.96
Toluene	3,69	2.46	2.19	2.19	1.92
Dichloroethane	3.33	2.32	2.22	2.19	1.92
Ethanol → water → dichloroethane	–	–	2.18	2.19	1.80
Dichloroethane → heptane	–	–	1.72	1.99	1.81
Ethanol → water → heptane	–	–	1.69	1.99	1.78

path leading to the swollen state. Thus, the swelling capacity with heptane does not depend on whether the polymer was taken in dry state or in the previously swollen (with dichloroethane or water) state. Swelling with water appears considerably smaller than that with other solvents, and differs also in that water does not wet immediately the hydrophobic material. It has to be wetted first by any precursor (acetone, alcohols, dioxane) and then washed with water to arrive at the equilibrium swelling level, which is independent of the precursor.

The ability of hypercrosslinked polymers to swell in non-solvents for polystyrene further depends on the rigidity of the network. Thus, structures crosslinked by trifunctional CMM exhibit the marked affinity towards hydrocarbons at noticeably smaller crosslinking degrees than polymers obtained with CMDP or XDC (Table 2). On the contrary, materials prepared with the flexible diphenylbutane-type reagent do not swell with hydrocarbons at

Table 2, Swelling capacity (ml/g) in hexane of hypercrosslinked polymers prepared from linear polystyrene

Degree of crosslinking	Crosslinking agent		
	CMM	MCDE	DPB
5	0.08	0.06	0.07
11	0.20	0.06	0.07
18	1.03	0.21	-
25	2.21	0.48	0.17
43	3.10	3.48	0.26
66	3.10	4.30	0.32
100	3.10	4.12	0.18

all, in spite of the fact that they contain tetra-
methylene fragments that could be expected to enhance
affinity to hydrocarbons.

Several additional factors were found to influence
swelling properties of hypercrosslinked polystyrene.
Thus, the swelling rises with the molecular weight of the
initial polystyrene falling (in the range from 300000 to
8800 examined) as shown in Fig. 3. Another factor is the
strain of polymeric chains subjected to crosslinking.
Linear polystyrene chains in solution exist in an
equilibrium non-strained conformation and produce gels of

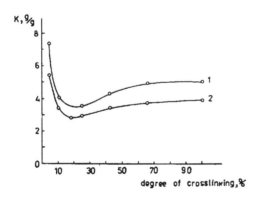

Fig. 3. Dependence of swelling in toluene of polymers
obtained by crosslinking with MCDE of polystyrene having
molecular weight of 8800 (1) and 300000 (2), as a
function of crosslinking degree.

high swelling ability. In the ultimately swollen granules
of styrene-DVB copolymers, on the contrary, some (elastically active) chains are strongly strained, and in this
pre-strained state they have to be incorporated into the
final hypercrosslinked network, which results in the definite diminishing of the equilibrium swelling [12]. Naturally, all other synthesis conditions in these comparative experiments should be kept constant, and, first of
all, the polymer-to-solvent ratio which influences the
polymer properties especially strongly [13].

Of the above discussed factors, the degree of network
crosslinking and thermodynamic quality of the solvent,
which modern swelling theories are based on, play a sub-
ordinate role, only.

When in dry state, hypercrosslinked polystyrene represent transparent materials of anomalously low density.
The total volume of voids amounts to 0.2-0.5 cm^3/g, as
calculated from the density or measured by adsorption of
nitrogen at -196° C (Table 3). Noteworthy, the latter method produces considerably higher values than the former,
which can only be explained by swelling phenomena of the
polymer in the nitrogen condensed. Naturally, the BET
adsorption theory should not be applied to such systems,
in spite of the fact that experimental data perfectly fit
corresponding equations to produce (apparent) inner specific surface area values of 1000-1200 m^2/g. Nevertheless, these data show that the polymers readily absorb
nitrogen and noble gases at low temperatures and behave
with this respect identically to porous materials having
inner surface area of that high magnitude.

Adsorption of gases by porous polymeric materials,
that were known thus far, takes place on the surface of
pores which can be easily detected by electronic microscopy or other methods revealing structural heterogeneity. Hypercrosslinked polystyrene, however, is transpa-

Table 3, Apparent specific inner surface area (S, m^2/g), apparent density (ρ, g/cm^3), and total volume of voids (W, cm^3/g) as determined from the density (W^P) or from nitrogen adsorption isotherms (W^{N_2}) for hypercrosslinked polymers prepared from linear polystyrene

Degree of crosslinking, %	Cross-agent	S	ρ	W^P	W^{N_2}
25	CMDP	0 to 100	0.86	0.27	—
43	CMDP	670	0.85	0.28	—
66	CMDP	800	0.81	0.34	—
80	CMDP	—	0.79	0.37	—
100	CMDP	1000	0.79	0.37	—
25	XDC	0 to 100	0.84	0.30	—
43	XDC	500	0.77	0.41	—
66	XDC	800	0.74	0.46	—
80	XDC	—	0.73	0.48	—
100	XDC	1000	—	—	—
25	MCDE	240	0.91	0.21	0.09
43	MCDE	650	0.80	0.36	0.46
66	MCDE	1000	0.75	0.44	0.54
80	MCDE	—	0.71	0.51	0.64
100	MCDE	1000	0.71	0.51	0.68

rent, and even electronic microscopy fails to reveal any pores or interfaces. Moreover, according to the latter technique, beads of styrene copolymers with 2 % DVB are completely deprived of any kind of inner supramolecular structure, both before and after their additional crosslinking, in spite of the fact that the latter procedure introduces 0.2 cm^3/g of voids and results in an apparent inner surface area of 1000 m^2/g.

This "structure-less" material was also examined using small-angle X-ray scattering technique. Evaluation of its scattering pattern can be described by the model of cylindrical inhomogeneities, 90-120 A^o in length and 25 to 30 A^o in diameter, when in dry state, or 15-20 A^o in diameter, when swollen with hexane [14]. If they were voids, their volume would amount to circa 3-5 % of the

volume of dry material and 10 % of that swollen in hexane
(provided that the density of the bulk polymer remains
close to that of polystyrene, 1.04 g/cm^3). This model does not correlate with the facts that the material incorporates almost 20 % voids when in dry state and takes up
to 1 ml/g of hexane on swelling. Evidently, major part of
voids does not scatter X-rays under low angels intensively enough, i. e., they cannot be represented by channels
and pores with any definitely organized walls.

Examining the swollen with chloroform hypercrosslinked
polystyrene using gel permeation chromatography, reveals
its full accessibility to test species of circa 15 A^o in
diameter. In this region, locates the formally calculated
very sharp maximum of pore size distribution. Polystyrene
coils of 50 to 60 A^o in diameter penetrate the gel material to negligible extent, only. These results, again,
show sufficient homogeneity of the network structure, its
marked difference from conventional porous two-phase polymeric structures.

Differently and very peculiar appears the morphology
of hypercrosslinked polystyrene prepared from the linear
dissolved polymer: spherical moieties of circa 120 A^o in
diameter are seen on cleaves of the material (for starting polystyrene of 300000 D molecular weight). The diameter of the spherical structures increases with the molecular weight of the linear polystyrene rising, but it
is totally independent of the degree of crosslinking in
the range of 5 to 100 % examined (note that the properties of the material change dramatically in this range).
The above and some additional experiments lead us to the
conclusion that the spherical moieties could correspond
to individual macromolecular coils in the initial polystyrene solution. This would imply, however, that the
coil in the concentrated solution, does not penetrate
significantly into its neighbour's coils and has the-

refore a much denser conformation than any predicted one. Crosslinking reaction merely fixates this structure of the concentrated solution [15].

STRUCTURE OF HYPERCROSSLINKED NETWORKS

None of the above mentioned physico-chemical methods could reveal any features of a two-phase nature or typical heterogeneity of the hypercrosslinked polystyrene material. Indeed, there seem to be no reasons for a real micro-phase separation of the system to occur during the synthesis of these polymers. First of all, introduction of cross-bridges shown in Fig. 1 does not affect the chemical nature of the polymer to any noticeable extent. Therefore, polymeric chains remain strongly solvated with dichloroethane from the beginning to the very end of the reaction. Secondly, when dissolved in dichloroethane, individual polystyrene coils of an average molecular weight of 300000 D begin to contact each other at the concentration of 1.5 %, and it would be logical to assume that at the concentration of 10 % used in the cross-linking procedure the overlapping of the coils already reaches a significant level that implies a rather equal distribution of polymer segments throughout the whole volume of the initial solution. Similarly, the cross-agent and the catalyst are distributed evenly in the reaction volume. Therefore, the reaction starts simultaneously in the whole mixture and leads to nearly statistic distribution of crosslinks in the gel formed. Obviously, a deviation from the statistics can only arise at a relatively advanced stages of the reaction where pending functional groups of the initially bifunctional reagent do not find immediately a second reaction partner. At these stages, however, the macromolecules are already involved into formation of a dense spatial network and they neither can segregate into a separate phase nor form supramolecular structures.

For these reasons, we seek to explain the unusual properties of hypercrosslinked polystyrene in terms of remarkable topology of its network rather than its supramolecular construction.

Starting from the suggestion that crosslinking concentrated polystyrene solution results in homogeneous statistically organized gel, we suppose its network to be a
system of mutually condensed and interpenetrating spatially nonplanar cycles that are formed by crosslinking
agents and chain segments confined between the branching
points. The contour length of the cycles depends on the
degree of crosslinking and the length of the crosslinking
agent molecule. At a formal 100 % degree of crosslinking,
all initial phenyl rings are involved into formation of
cross-bridges, and the network appears to be built of
smallest cycles, especially when the cross agent (MCDE or
DMF) merely contributes a methylene group. Fig. 4 illustrates the structure of such smallest possible cycles
made up of two (or three) pairs of neighbouring phenyl
residues belonging to two (or three) different chains or
chain sections and two (or three) methylene links. The
first cycle appears to be strained, and its formation
probability should be considered rather small. Formation
of larger cycles is more probable, of course.

Two evident conclusions can be made from the above
structure of the network cycles.
(i) The cycles are built of sufficient number of carbon-
to-carbon bonds, that should allow considerable changes
in their conformation. In the network, however, every
cycle is interconnected, mutually condensed with a large
number of neighbouring cycles and cannot behave independently of them. Nevertheless, a cooperative conformational rearrangement of a large assemble of cycles should
make possible considerable changes of the network volume

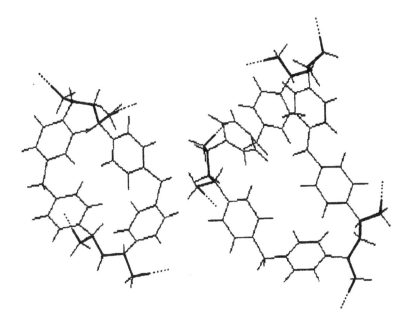

Fig. 4. Schematic presentation of smallest possible cycles composing a hypercrosslinked polystyrene network.

on drying or swelling of the polymer. Expansion of the network on swelling should be facilitated by the remarkably lower degree of interpenetration of the network cycles in hypercrosslinked polystyrene as compared to those in conventional densely crosslinked styrene-DVB copolymers or epoxy resins, since in the former case, the cycles were closed in the presence of very large amounts of a good solvent. At least, attaining the equilibrium swollen state that incorporates these amounts of the solvent should proceed easily and cause no additional strains of the network.

(ii) Sterical hindrances and considerable inner strains of the network cycles should rapidly grow on removing the solvent and contracting the gel volume. It is only natural that attaining a dense packing of polymeric chains appears impossible, and many small voids should remain in

the network, which markedly diminish the density of dry
polymer, but do not scatter X-rays intensively.

Internal strains in the transparent granules of dry
hypercrosslinked polymer are clearly seen in a polari-
zation microscope in the form of so-called Maltese cross.
On swelling, these strains relax to a considerable extent, and, in our opinion [15], this relaxation tendency
causes the ability of hypercrosslinked polymers to swell
in any liquid or gaseous media.

REFERENCES

[1] Malinsky, J., Klaban, J. and Dušek, K., J. Macromol.
     Sci., A-5, 1071-1085 (1971)
[2] Millar, J.R., Smith, D.G., Marr, W.E. and Kressman,
     T.R.E., J. Chem. Soc., 1963, 218-225
[3] Rempp, P., Herz, J., Hild, G. and Picot, C.,   Pure
     Appl. Chem., 43, 77-96 (1975)
[4]   Anderson, R.E., Ind. Eng. Chem., Prod. Res.
     Develop., 3, 83-85 (1964)
[5] Dušek, K., J. Polymer. Sci., B-3, 209-216 (1965)
[6] Saidl, J., Malinsky, J., Dušek, K. and Heitz, W.,
     Adv. Polymer. Sci., 5, 111-213 (1967)
[7] Samsonov, G.V. and Kuznetsova, N.P., Adv. Polymer.
     Sci., 1991, in press; Samsonov, G.V., Pisarev,
     O.A. and Muravieva T.D., Vysokomol. Soedyn., 28-B,
     262-264 (1986)
[8] Howard, G.J. and Midgley C.A., J. Appl. Polymer.
     Sci., 26, 3845-3870 (1981)
[9] Tarakanova, E.E. and Volodavets, I.N., in Problems
     of Physico-Chemical Mechanics of Fibrous and
     Porous Dispersed Structures and Materials, Zinatne,
     Riga, 1967, 95-101
[10] Dušek, K., Saidl, J. and Malinsky, J.,   Collect.
     Czech. Chem. Commun., 32, 2766-2778 (1967)
[11] Tsyurupa, M.P., Lalaev, V.V., Belchich, L.A.   and
     Davankov V.A., Vysokomol. Soedyn., 28-A, 591-595
     (1986)
[12] Tsyurupa, M.P., Andreeva, A.I. and Davankov, V.A.,
     Angew. Makromol. Chem., 70, 179-187 (1978)
[13] Davankov, V.A., Rogozhin, S,V. and Tsyurupa, M.P.,
     Angew. Makromol. Chem., 32, 145-151 (1973)
[14] Authors are grateful to Dr. A.N. Oserin for measu-
     ring X-ray scattering and for useful discussions.
[15] Davankov, V.A. and Tsyurupa M.P.,   Reactive
     Polymers, 13, 27-42 (1990)

Polymer Networks '91 pp. 183-198
Dosek and Kuchanov (Eds)
© VSP 1992

Physical networks of biopolymers

Simon B. Ross-Murphy

Cavendish Laboratory, University of Cambridge, Madingley Road, Cambridge CB3 0HE, UK

Abstract: Polymer networks can be divided formally into two classes, chemically cross-linked materials (including bulk elastomers), and "entanglement networks". Between these two categories fall a number of systems, which consist of chains "physically" cross-linked into networks. Physical gels can be formed from both synthetic and naturally occurring polymers. Amongst the latter are gelatin, the seaweed and plant polysaccharides such as agarose, the carrageenans and pectin, starches and cellulose derivatives, globular protein gels, formed either by heating or by change of chemical conditions (pH etc.) and fibrillar gels of actin and myosin. This article discusses the structure and mechanical properties of these gels, and the nature of the cross-links in the above increasing order of structural complexity. A difficulty in their characterisation is that such systems are often under kinetic rather than thermodynamic control.

COVALENT AND ENTANGLEMENT NETWORKS

Polymer gels or networks can be divided into two main classes, chemically cross-linked materials (including bulk elastomers), and "entanglement networks". Covalently cross-linked materials are formed by a variety of routes including cross-linking high molecular weight linear chains, either chemically or by radiation, by end-linking reactant chains with a branching unit, or by step-addition polymerisation of oligomeric multi-functional precursors, and are discussed elsewhere in this volume. They are true macromolecules, where the molecular weight is nominally infinite, and they therefore possess an infinite relaxation time and an equilibrium modulus (Ref.1).

Note a. Current Address: Biomolecular Sciences Division, King's College, Kensington Campus, Campden Hill Road, Kensington, London W8 7AH

Entanglement networks are formed by the topological interaction of polymer chains, either in the melt or in solution when the product of concentration and molecular weight becomes greater than some critical molecular weight M_c (Refs.1,2). In this case they exhibit a pronounced plateau modulus, which extends to lower and lower frequencies as the concentration/molecular weight increases. They therefore behave as "pseudo gels" at frequencies higher (timescales shorter) than the "lifetime" of the topological entanglements. Much progress has been made in describing the dynamics of such systems using "tube" models. For example the Doi-Edwards theory shows how the relaxation behaviour of linear chains depends on M_r^3, where M_r is molecular weight. The theory formulates a constitutive equation for entangled systems, whose molecular parameters may be verified independently (Ref.2).

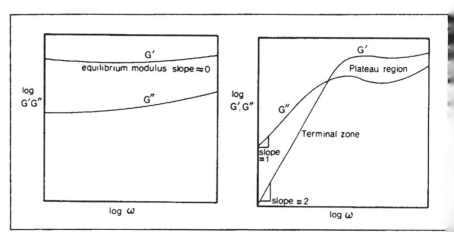

Fig.1.   Expected dynamic mechanical spectra for the real (G') and imaginary (G") parts of the shear storage modulus for (left) a covalently cross-linked network and (right) an entanglement network system (pseudo-gel).

The differences these two classes of networks can be most clearly seen by viscoelastic particularly dynamic mechanical measurements (Ref.1). A small deformation, oscillatory strain (for "soft" solids and fluids this usually a shear strain, γ) of frequency is applied to the material, and the real and imaginary parts, G' and G", of the comple shear modulus are measured as a function of ω, γ, temperature T etc. Typical spectr are illustrated in Fig.1; for the entanglement networks, as the frequency is decrease there is a "cross-over" in G' and G''. At very low frequencies, in the "terminal zone" they flow as high viscosity liquids, but the precise behaviour clearly depends upon C, M_r and T.

The corresponding spectrum for the cross-linked system depends upon whether or not CM_r for the system before cross-linking was above or below M_e. In the case where $CM_r < M_e$, corresponding particularly to step-addition polymerisation, and the degree of cross-linking is substantial, the cross-linked networks show little effect of entanglements, and the traces of (log) G' and G'' vs. (log) ω are parallel, and with a slope close to zero; tan δ, the loss factor ($= G''/G'$) is typically < 0.05. Such a mechanical spectrum demonstrates that the corresponding relaxation spectrum is extremely broad, reflecting contributions from the many "path lengths" in a branched network. Close to the gel point the frequency dependence is more pronounced, and actually at the gel point work by Winter and co-workers, particularly on PDMS networks, but also more recently on physical networks (Ref.4), has shown that there is power law behaviour, i.e. G' and G'' are both proportional to ω^x, and the exponent x is constant over a range of frequencies (experimentally 3-4 decades).

A further, more classical discrimination between gels and pseudo-gels can be made by adding a large excess of solvent. Entanglement network systems will dissolve to form a more dilute polymer solution, whereas in covalent systems whilst the sol fraction will dissolve, the gel fraction can only swell.

PHYSICAL GELS

For both of the above classes of network the behaviour can be said to be largely understood, whilst controversies still exist, for example regarding the entanglement contribution in cross-linked networks, most discrepancies between theory and experiment are at the refinable level. However, quite a number of systems exist which consist of chains "physically" cross-linked into networks, the cross-links themselves being of small but finite energy and/or lifetime. These possess some properties of both categories, and are called physical gels[b], a description which includes both biological

---

Note b. The term 'physical gel', which appears to have been introduced by de Gennes[3], is often assumed to imply thermoreversibility, which is by no means the case with every system included here. Instead we will consider the term to incorporate any non-covalently cross-linked system.

and synthetic polymers (Ref.5). The presence of non-covalent cross-links complicates any physical description of the network properties enormously, because their number and position can, and does, fluctuate with time and temperature.

In many cases the nature of the cross-links themselves is not known unambiguously, often involving such disparate forces as, for example, Coulombic, dipole-dipole, van der Waals, charge transfer and hydrophobic and hydrogen bonding interactions. For biopolymer gels, in particular, non-covalent cross-links are formed by one or more of the mechanisms listed above, almost invariably combined with more specific and complex mechanisms involving extended quasi-crystalline "junction zones" of known ordered secondary structure eg. multiple helices, ion mediated "egg box" etc. (Ref.6). Typically there is a specific, and often intricate, hierarchy of arrangements, a situation which is more familiar to molecular biologists than to polymer scientists. For example gels formed from the marine polysaccharide carrageenan, are generally accepted to involve intermolecular double helix formation (Ref.7). Here the existence of "rogue residues", saccharide units which are in the wrong conformation to allow the helix to propagate, and which therefore terminate the sequence, was established chemically. This ensures that, rather than isolated helix pairs being formed, each chain can share portions of ordered helical structure with at least two other chains, an essential condition for branching and subsequent gel formation.

Physical description of such networks is intrinsically rather difficult, because the potential number of junction zones per primary chain, the "functionality", f, and the extent (molecular weight) of the junction along the chain profile can be estimated only indirectly. In many cases there is a subsequent lateral aggregation of chains, after the initial contact. These factors must influence the actual number of physical cross-links and, consequently, the modulus of the final gel. Since this modulus will usually reflect both entropic (rubber-like elasticity) and enthalpic contributions, an *a priori* description of the modulus and mechanical response of these materials is bound to involve number of approximations.

STRONG AND WEAK GELS

A further discrimination can be made between so-called "strong" and "weak" physical gels (Ref.8). The latter can be further divided into "rheological" and "thermodynamic"

weak gels, in the latter case the energy per cross-link is only a few kT so they can be
cleaved by thermal fluctuation, only to reform. Many thermoreversible biopolymer gels

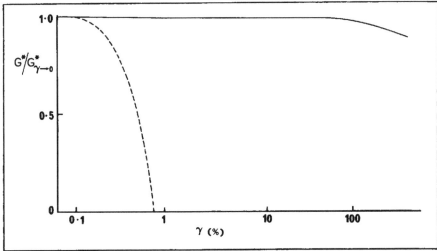

Fig.2.   Typical strain dependence of a strong (solid) and a weak (dotted) gel.

can only be melted by passing through an order-disorder transition (with associated
mid-point transition temperature). Both strong and (rheological) weak gels respond as
solids at small deformations, but whereas the former, eg. gelatin are also solids at
larger deformation, the latter are really structured fluids, so they flow almost at liquids
at large deformations[c]. Rheologically this means that whilst the mechanical spectrum
within the linear (stress proportional to strain) viscoelastic regime is still as illustrated
in Fig.1 (left), the maximum linear strain, γ_0, is very different (Fig.2). In the former
case we usually say that $\gamma_0 > \approx 20\%$, for weak gels it can be up to 1000 times smaller.
Colloidal and particulate networks are often of this type. In the present text we mainly
restrict ourselves to strong gels, and only to a cursory discussion of most of these.
Discussion will focus chiefly on gelatin (cold set) and globular protein (heat set) gels.

Note c. This distinction must be somewhat arbitrary since if the non-covalent
cross-links can be ruptured by mechanical perturbation, they must have bond
energies of only a few kT units, and therefore may also "melt" on heating.
Ultimately such discussion serves only to focus attention on the inadequacy of
using the term "weak gel" to describe a range of different phenomena.

NETWORKS FROM DISORDERED POLYMERS

A fundamental distinction can be made between those systems which in forming gels are transformed from essentially "disordered" random coil biopolymers (although even in the disordered conformation, in terms of say persistence length, these are sometimes "stiff" compared to synthetic polymers) to a partly ordered state (eg. by undergoing a coil-helix transition), and those systems which form, and are maintained in an essentially ordered state, such as gels formed from globular proteins. The former usually gel *in vitro* by mechanisms in which they partially renature to their *in vivo* state.

GELATIN GELS

The paradigm for biopolymer gels is gelatin and indeed the term gel (attributed to Thomas Graham) is originated from it. Gelatin(e) is a proteinaceous material derived by hydrolytic degradation of collagen, the principal protein component of white fibrous connective tissue (skin, tendon, bone etc.) with as fundamental molecular unit the tropocollagen rod. The latter is a triple helical structure composed of three separate polypeptide chains (total molecular weight $\approx$ 330,000, persistence length $\approx$ 180 nm). Gelatins normally dissolve in warm water ($> \approx 40C$) and above this temperature the polypeptide exists as flexible single coils. On recooling, transparent gels are formed (provided the concentration is greater than some critical concentration, C_0, typically 0.4 to 1.0%). The gels contain extended physical cross-links or "junction zones" formed by a partial reversion to "ordered" triple helical collagen-like sequences, separated along the chain contour by peptide residues in the "disordered" conformation. It was presumed that the gelatin triple helices involved three separate intermolecularly wound peptide chains, as in the original tropocollagen helix, and that each chain participates in several such junction zones. The topological consequences of this on subsequent helix formation and gelation are very significant, and perhaps for that reason are scarcely ever discussed! However, on the basis of the concentration dependent order of kinetics observed by optical rotation (OR), a technique which directly monitors the proportion of residues in the triple helical conformation, it has been proposed that helix nucleation is a bimolecular process, involving an intramolecular ß-turn and another gelatin macromolecule (Refs.9,10). When a third segment meets a "kink" with the correct orientation, a triple helix is nucleated. Fig. 3 illustrates the alternative hypotheses. The nuclei are, of course, not stable unless a critical minimum size is reached, a size which depends on temperature, and corresponds to the balance between an initial loss

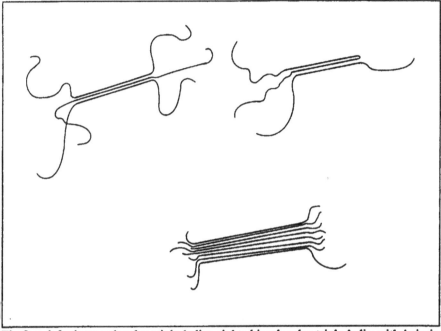

Fig.3.   left: intermolecular triple helix; right: bimolecular triple helix with hairpin bend (represented as parallel lines); bottom: side-by side aggregation of helices as originally suggested to describe gelatin junction zones.

of entropy and the enthalpic stabilisation due to helix formation. This size has usually been estimated from observations on very low molecular weight gelatin samples, since these cannot renature at all, and the evidence seems to favour a length of between 20 and 40 peptides (Ref.10). As for the precise mechanism of helix growth it is thought that the coil to triple helix propagation rate is limited by the presence of *cis*-proline residues in the backbone. The subsequent reversion of these to the *trans-* form allows the helix to propagate only slowly, the overall growth rate is typically 4-6 orders of magnitude less than for double helical systems, such as the carrageenans (see below).

Until recently it was postulated that initial helix formation was followed by substantial lateral aggregation leading to extended "quasi-crystalline" junctions, as described historically in the fringed micelle model of polymer crystallization. However since long time measurements of OR increase slowly, but apparently without limit (even when plotted on a log time axis) then the proportion of residues in the ordered helical conformation must also be increasing. This suggests a considerable degree of conformational flexibility, even post-gel, and is rather unlikely to occur if the junction zones are formed of rigid crystallites. More direct evidence against this comes from SANS by Djabourov and her co-workers (Ref.12). They found that in the sol state the cross sectional chain radius, R_c was 0.32 +/- 0.1 nm, a value in good agreement with the calculated side-group extension of a collagen chain, whereas in a relatively concentrated gel (5%) R_c was 0.43 +/- 0.1 nm, not much greater than this. It is now thought that the junction zones consist of (largely) isolated triple helices (see eg. Ref.13).

MARINE POLYSACCHARIDES

The most important of these forming gels are (ι- and κ-) carrageenan, agar(ose), and the alginates. Much evidence suggests that the first two form thermoreversible gels by an extension of the gelatin mechanism, and, although some details are still disputed the general principles are as below. On heating above the helix-coil transition temperature (for the charged carrageenans this depends crucially on ionic strength and cation species but typically in the range 20-50C), they disorder. On recooling they partly revert to a double helix (Refs.6,7), for agarose there is then substantial side-by side aggregation (confirmed by measurement of R_c by SAXS) (Ref.14). For the carrag eenans gelation is known to depend crucially on the cations present, for Na^+ little is seen, whereas high modulus gels are formed for example with K^+ and Ca^{2+}. This is consistent with the "domain" model proposed by Morris and co-workers (Ref.6), in which junction zone formation involves ion mediated aggregation of double helical regions. The precise details of network formation in these systems is still being actively researched.

Alginate gels are not thermoreversible, in fact they appear heat stable up to >100C and their formation can only be induced by certain, specifically divalent, cations

Alginates are anionic block copolymers composed of two very similar saccharide units, guluronate (G) and mannuronate (M). If Ca^{2+} ions are introduced into a solution of sodium alginate, gelation occurs extremely rapidly. Gelation is induced by specific ion binding accompanied by conformational change, and circular dichroism evidence implies that Ca ions bind cooperatively to G blocks. In one model the junction zones involve two chains and chelated ions giving the so-called "egg box" structure (Ref.15).

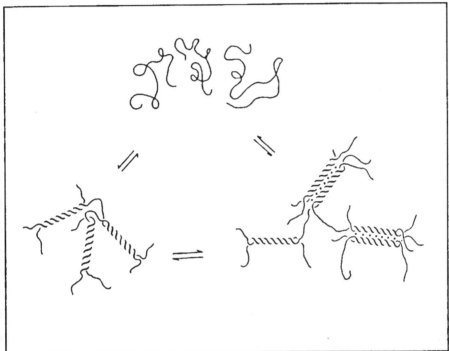

Fig.4.　Domain model of carrageenan gelation (after Morris and coworkers - Ref.6).

PLANT POLYSACCHARIDES

Pectin and starch gels are the most important members of this group, the former consist predominantly of sequences of galacturonic acid residues (which are quite similar to the G units in alginate), with occasional interruptions by rhamnose residues. At least some of the carboxyl groups are methyl esterified, the precise distribution depending upon the plant source and age. Reasonably in view of their structural similarity, pectins of low degree of esterification behave like alginates, and gel with divalent ions. The more esterified materials gel under conditions of low pH and decreased water active, ie.

where intermolecular electrostatic repulsions are reduced; in this case the junction zones are thermoreversible at say 40C.

Starch consists of two different polysaccharides, one, predominantly linear, being amylose and the other branched, but otherwise structurally analogous, amylopectin. On a weight basis amylopectin usually predominates (>70%), but the precise proportion of the two polymers, their chain length and branching frequency depends upon the source (potato, wheat, rice, tapioca etc.) The polymers themselves are laid down in an ordered semi-crystalline supramolecular granule ($\approx$ 2-100μm), and on heating granules swell and rupture (at temperatures >$\approx$60C). After this it is believed that the amylose leaches out, and at concentration >say 20% the mixture of granular residues and amylose forms a viscoelastic paste. On cooling this sets up, and the result has been described as a composite of amylose gel filled with swollen granules (Ref.16). This is, however, by no means the whole picture since amylopectin solutions can also gel, and much of the subtlety of starch behaviour is undoubtedly related to the limited compatibility, and mutual gelation of the two polymeric components.

MICROBIAL POLYSACCHARIDES

A number of polysaccharides of interest occur outside the cells of certain cultured microbes, either covalently attached or secreted into the growth media. These are the microbial exopolysaccharides, and over the past few years a great number of these have described (Ref.17). At the moment, on a volume production basis, the two major members of this group are gellan, an anionic polysaccharide produced by *Auromonas elodea*, and xanthan, also anionic, from *Xanthomonas campestris*. Gellan has a complex tetrasaccharide repeat unit, and gels in the presence of multivalent cations, via a double helical intermediate, in a way analogous to the gelling carrageenans. The bulk mechanical properties are sensitive to the degree of acylation of the chain. Xanthan which has a pentasaccharide sequence, forms strong gels only under extreme conditions. It is, however, the archetypal weak gel structurant, and has been employed in a number of industries because of its rather unusual rheological properties (Ref.8).

NETWORKS FROM GLOBULAR AND ROD-LIKE BIOPOLYMERS

Almost all of this group of materials are formed from animal and vegetable proteins

In some cases the resultant gels involve, at least partial, denaturation which does not occur *in vivo* (heat set proteins) whilst in others the biological function of the protein is to gel under certain physiological conditions (blood clotting = fibrin network formation). Distinction must also be made between established gel systems and solutions of entangled rods; there are several cases in the literature where these are referred to as gels.

GLOBULAR PROTEINS FORMING BRANCHED NETWORKS

Many globular proteins can form gels, above say 5% concentration, just by heating. Perhaps the most familiar example of this is the boiling of an egg (essentially gelation of ovalbumin), but similar gels can be formed by heating eg. serum albumins (SA), chymotrypsin, globins, whey and vegetable (soy) proteins. Much of the published work has been restricted to, particularly bovine (B) SA, since pure samples of this can be obtained relatively cheaply. There are many published data on the rheological properties of all the systems above, but in many cases comparatively crude samples have been used. As an example whey protein usually consists of mixtures of lactalbumin, lactoglobulin and smaller amounts of caseins.

For systems heated to not much greater than the protein unfolding temperature (≈70C) it was first thought that the mechanism of gelation was somewhat akin to that of gelatin, heating (denaturing) producing a massive conformational change which converted serum albumin - an ellipsoidal globule approximately 6nm by 4nm - into a random coil polypeptide, which then partly refolded intermolecularly to form cross-links involving the peptide ß-sheet conformation (Ref.18). Studies by Barbu and Joly (Ref.19) and by Kratochvíl and co-workers (Ref.20) on horse and human SA respectively, and using a variety of techniques led to an alternative hypothesis. The conformation of the corpuscular protein is only slightly perturbed, and subsequent work using electron microscopy, X-ray scattering and spectroscopic techniques has confirmed this (Ref.21).

It appears that denaturation partially disrupts the protein without modifying the overall shape very significantly, but exposes some intraglobular hydrophobic residues. At low enough concentrations these can refold all but reversibly, but above a certain concentration there is competition between intra- and intermolecular ß-sheet formation. If the latter predominates, gels are formed which are fibrillar, and whose fibrils are approximately 1-2x the width of the original globule. The balance between linear and assemblies by respectively addition of salt and action of other proteins. The polymeric actin fibril is composed of two helical strands wound around one another, whereas that formed from tubulin "monomers" is a hollow cylinder. The fibril length can be very great (up to 30μm for actin, DP $\approx$ 10000), and these fibrils then form very weak networks. Treatment with a so-called actin binding protein (ABF) provides covalent cross-links, whilst another protein gelsolin cuts actin filaments producing a drastic reduction in gel modulus (Ref.24). Such actin networks are crucial to the self-regulating mechanical behaviour of the intra-cellular cytoplasm.

Fibrinogen is a relatively compact protein, which can also polymerise under physiological conditions to form very fine stranded fibrillar networks. The rheological properties of these have been extensively investigated by Ferry and co-workers, and they have demonstrated how the mesh-size of the network can be altered by changes in pH and I. Ca^{2+} in the presence of the serum factor fibrinoligase produces covalent bonds (Ref.25).

CASEIN GELS

Casein (milk) gels are important technologically because they form the basis of cheese, yoghurt and other similar products. The term casein, itself, describes a number of different proteins (α_{s1}, α_{s2}, ß and κ-caseins), which occur in milk as roughly spherical but highly voluminous micelles, typical DP $\approx$ 10000, particularly stabilized by colloidal calcium phosphate. Treating whole milk with an enzyme chymosin (rennet) is believed to cleave away the κ-casein which exists on the "outside" of the micelle, producing a coagulate (curd), which is separated from the remaining liquor (whey), a solution of whey proteins. Many detailed studies of casein aggregation have been carried out, although most often using the methods of classical colloid science (Ref.26).

Casein aggregation can also be induced by heat ($>$140C), Ca^{2+} ions, and acid treatment; the lowering of pH by bacterial action is the basis of yoghurt making. Recent work by Walstra, van Vliet and co-workers has employed a fractal aggregation model to describe the mechanical properties of gels formed both by rennet and acid treatment (Ref.27).

branched chain growth depends upon both pH and ionic strength I, and a range of "gels" can be prepared ranging from transparent through translucent to biphasic gels and turbid coagulates without macroscopic strength. In fact boiled egg white is only white because of the concentration of salt in the ovalbumin solution, suitable dialysis can produce a transparent "white".

For globular gels formed by heating it is rather unlikely that thermodynamic equilibrium has been achieved but, at least experimentally, "phase" diagrams can be constructed showing the boundary between sols, clear gels and turbid gels as a function of protein concentration, pH and I; a similar mapping have been performed for gel modulus (Refs.21,22). Nevertheless there are few detailed similarities between different proteins; even BSA and ovalbumin, which seem to behave quite analogously, and which apparently form completely compatible co-gels, differ at the level of protein secondary structure. BSA gels have slightly less α-helix than in the sol, but more residues in the β-sheet and disordered conformations. For ovalbumin gels the proportion of β-sheet is markedly greater. For both, however, the proportion of β-sheet in the gel is largely independent of the degree of cross-linking, intimating that the intraglobular β-conformation is not the only factor involved. Prolonged heating at temperatures $>85C$ produces a more drastic change, and some intermolecular covalent disulphide bonds are formed; these gels can no longer be regarded as merely physical networks.

Protein gels can also be produced by other means of denaturation than simply heat, including treatment with non-solvents (alcohols), or "hydrogen bond breakers" (urea). Subjecting ovalbumin solutions to both urea and heat ($>80C$) apparently produces very substantial peptide unfolding and, in this case, the mechanism of gelation may be closer to the gelatin-like picture originally proposed (Ref.23). Gels can also be produced by enzymic means, crucially important in the antibody-antigen reaction, and also in the process of red blood cell agglutination.

GLOBULAR PROTEINS FORMING ORDERED FIBROUS ASSEMBLIES

In this category we consider specifically actin, tubulin and fibrin networks. The first two of these are globular proteins ($\approx$6nm) which can form very specific rod-like

NETWORKS FROM ROD-LIKE PROTEINS

The main system of interest here is myosin. This, the principal protein of muscle, is a rod-like macromolecule ≈ 150nm in length, one end of which is formed into two lobes - the head group. Myosin gels can be formed by heating to >60C, and the specific properties again depend on pH and I. Depending upon pH gels are formed either from myosin monomer (pH >6) or from filamentous aggregates (Ref.28). As one might expect the former are more heterogeneous in structure. Gels formed from myosin filaments involve aggregation of the myosin head groups, distributed on the "outside" of the filaments. Collagen is also a rod polymer, but normally forms gels, only after its structural degradation, to give gelatin, as described earlier.

CONCLUSION

Although largely a descriptive review, it is hoped this article has helped to clarify some of the issues, and areas of progress in biopolymer gelation for workers more familiar with synthetic polymers. The area is one where there is increasing interest, and still many unsolved problems.

ACKNOWLEDGEMENTS

The Author thanks Prof V. Kabanov, Prof Yu. Godovsky and the Polymer Council, USSR Academy of Sciences for their hospitality and the opportunity to participate in this Symposium. He is grateful to the Royal Society, and the SERC for financial support during the course of this work and Prof Sir Sam Edwards for his extended hospitality.

REFERENCES

(1)   J.D. Ferry, Viscoelastic Properties of Polymers 3rd Edition , John Wiley, New York (N.Y.) 1980

(2)   M. Doi, S.F. Edwards, The Theory of Polymer Dynamics Clarendon Press, Oxford, 1986

(3)   P.-G. de Gennes, Scaling Concepts in Polymer Physics Cornell University Press, Ithaca (N.Y.), 1979

(4)   Y.G Lin, D.T. Mallin, J.C.W. Chien and H.H. Winter, Macromolecules 24, 851 (1991)

(5) W. Burchard, S.B. Ross-Murphy (Eds.), Physical Networks - Polymers and Gels Elsevier Applied Science, London, 1990

(6) A.H. Clark, S.B. Ross-Murphy, Adv.Polym.Sci. 83, 57 (1987)

(7) E.R. Morris, D.A. Rees, G. Robinson, J.Mol.Biol. 138, 49 (1980)

(8) R.K. Richardson, S.B. Ross-Murphy, Int.J.Biol.Macromol. 9, 257 (1987)

(9) J.P. Busnel, S.M. Clegg, E.R. Morris, in Gums and Stabilizers for the Food Industry IV (Eds. G.O. Phillips, D.J. Wedlock, P.A. Williams) IRL Press, Oxford, 1988, p.105

(10) J.P. Busnel, E.R. Morris, S.B. Ross-Murphy, Int.J.Biol.Macromol. 11, 119 (1989)

(11) M. Djabourov, J. Maquet, H. Theveneau, J. Leblond, P. Papon, Brit.Polymer J. 11, 169 (1985)

(12) T. Herning, M. Djabourov, J. Leblond, G. Takerkart, Polymer in press

(13) S.B. Ross-Murphy, Polymer submitted

(14) M. Djabourov, A.H. Clark, D.J Rowlands, S.B. Ross-Murphy, Macromolecules 22, 180 (1989)

(15) E.R. Morris, D.A. Rees, D. Thom, J. Boyd, Carbohydr.Res. 66, 145 (1978)

(16) M.J. Miles, V.J. Morris, P.D. Orford, S.G. Ring, Carbohydr.Res. 135, 257 (1985)

(17) V.J. Morris and M.J. Miles, Int.J.Biol.Macromol. 8, 342 (1986)

(18) J.D. Ferry, Adv.Protein Chem. 4, 1 (1948)

(19) E. Barbu, M. Joly, Farad.Discuss.Chem.Soc. 13, 77 (1953)

(20) P. Kratochvíl, P. Munk, P. Bartl, Coll.Czech.Chem.Commun. 26, 945 (1961)

(21) A.H. Clark, C.D. Lee-Tuffnell, in Functional Properties of Food Macromolecules (J.R. Mitchell, D.A. Ledward Eds.) Elsevier/Applied Science, Barking U.K., 1986 p.203

(22) R.K. Richardson, S.B. Ross-Murphy, Brit.Polym.J. 13, 11 (1981)

(23) F.S.M. van Kleef, J.V. Boskamp, M. van den Tempel, Biopolymers 17, 225 (1978)

(24) P.A. Janmey, S. Hvidt, J. Lamb, T.P. Stossel, Nature 345, 89 (1990)

(25) P.A. Janmey, E.J. Amis, J.D. Ferry, J.Rheol. 27, 135 (1983)

(26) D.G. Dalgleish, J.Dairy Res. 50, 331 (1983)

(27) P. Walstra, T. van Vliet, in Food Polymers, Gels and Colloids (E. Dickinson, Ed.) R.S.C. London, 1991 p.369

(28) B. Egelansdal, K. Fretheim, K. Samejima, J.Sci. Food Agric. 37, 915 (1986)

T - #0062 - 160425 - C0 - 240/160/11 [13] - CB - 9789067641456 - Gloss Lamination